CHEESEMAKING MADE EASY

by Ricki and Robert Carroll

A Garden Way Publishing Book

Storey Communications, Inc.
Pownal, Vermont 05261

Illustrations by Linda L. Taylor
Photos by Tommy Olof Elder

Garden Way Publishing was founded in 1973 as part of the Garden Way Incorporated Group of Companies, dedicated to bringing gardening information and equipment to as many people as possible. Today the name "Garden Way Publishing" is licensed to Storey Communications, Inc., in Pownal, Vermont. For a complete list of Garden Way Publishing titles call 1-800-827-8673. Garden Way Incorporated manufactures products in Troy, New York, under the TROY-BILT® brand including garden tillers, chipper/shredders, mulching mowers, sicklebar mowers, and tractors. For information on any Garden Way Incorporated product, please call 1-800-345-4454.

Printed in the United States by Capital City Press
Sixteenth Printing, December 1993

Library of Congress Cataloging-in-Publication Data

Carroll, Ricki.
Cheesemaking made easy.

Includes index.
1. Cheese. 2. Cheese — Varieties. I. Carroll, Robert.
II. Title.
SF271.C37 1982 637'.3 82-9300
ISBN 0-88266-267-8 AACR2

CONTENTS

Acknowledgements v

Preface vii

INTRODUCTION 1

EQUIPMENT YOU WILL NEED 5

INGREDIENTS AND HOW TO USE THEM 11

SOFT CHEESES 27

HARD CHEESES 47

WHEY CHEESES 87

GOAT'S MILK CHEESES 93

BACTERIAL AND MOLD-RIPENED CHEESES 101

QUICK CHEESES 119

GLOSSARY 121

Trouble-Shooting Chart 125

Suppliers of Cheesemaking Equipment 129

Newsletters 129

INDEX 131

Acknowledgements

We wish to thank the following individuals for their help in producing this book: Kate Spencer, for all the hours of typing and editing and deciphering our notes; Linda Taylor, for her wonderful illustrations; Tommy Elder, for his excellent photographs; Rodney Wheeler and his mother of Dunchideock, England, for the wealth of information they provided us on English hard cheeses; Dr. Donald Irvine of the University of Guelph, for providing us with techniques for the making of Blue cheeses; Sister Mary-Pia, Sister Christine, and Sister Assumpta from the Benedictine Monastery of Mont St. Laurier, Canada, for their gracious hospitality and the wealth of information they shared with us on goat cheesemaking and the making of Camembert cheeses; Dr. Brochu of the Institute Rossell of Montreal, Canada, for his technical advice on cheesemaking; Jean-Claude Le Jaouen of the Institute Ovin et Caprine, Paris, France, for the technical information he provided us for the making of goat cheeses; Joop Rademaker of Utrecht, Holland, for information he provided us on the making of Gouda cheese; and to all the readers of our newsletter *Cheesemakers' Journal* who have shared with us recipes and advice on home cheesemaking.

Preface

Our cheesemaking started years ago when a friend of ours talked us into purchasing two dairy goats. We soon found ourselves inundated with excess milk and began to fear that our chickens would start to lay curdled eggs if we fed them any more milk. Attempts to purchase home cheesemaking equipment led us into starting our own company, New England Cheesemaking Supply Company. We eventually acquired a Jersey cow and found ourselves making a variety of cheeses from both goat's milk and cow's milk.

This book is a result of experiences during the past years of our home cheesemaking. The recipes are for amounts of milk ranging from one quart to two gallons. There are sixty recipes in this book and for those who are just becoming acquainted with home cheesemaking we suggest that you begin with the soft cheeses. Lemon cheese, Queso Blanco, lactic cheese (acid curd cheese), and the cottage cheeses are all excellent to begin with, for these are all delicious and easy cheeses to produce and require a minimum of investment in equipment.

Our purpose in writing this book is to present the home cheesemaker with the basic principles and guidelines for making cheese. If in the course of your cheesemaking you come across a problem whose solution escapes you, please do not hesitate to contact us here in Ashfield for advice.

VINTAGE
MATEUS
ROSÉ

INTRODUCTION

Those of us who remember our first attempts at breadmaking look back with an indulgent smile upon the bowl of sticky, disobedient dough that clung to everything it touched and defied our inexperienced hands at every move. The indulgent smile is possible because we also remember that we finally beat it into submission and that it did everything the book promised it would — such as kneading into a wonderful, elastic mass, rising up gloriously under its dampened cloth cover, and filling the house, one hour later, with one of life's most heavenly smells.

No one seems to know why the art of breadmaking fled the factories and resettled with delicious comfort in our homes so far ahead of the art of cheesemaking. Bread and wine and cheese, besides being (collectively and singly) among life's greatest pleasures, share another common bond. They are all produced by a process of fermentation. Bread rises in the pan and wine matures its heady excellence from the action of yeast and sugar. Cheese, as if by magic, emerges from a kettle of milk because of the action of bacteria upon milk sugar. They solidify the casein in the milk and give it keeping qualities so it can be aged and cured to produce the incredible spectrum of flavors cheesemakers have devised.

There's a Trick to It

Just as there's a knack to handling bread dough, there are tricks to turning milk into cheese. You might have learned to make bread by watching your mother. Today, however, it's one mother in a million who can be counted on to hand down the techniques of cheesemaking to the next generation.

So that's what this book is designed to do — to keep alive the art and the pleasures of cheesemaking, and to stand by and hold your hand when the going gets rough, or the curds won't set, or your cheese turns monstrous and bloated, and you can't bring yourself to taste it.

If you follow directions carefully (especially the ones about cleanliness), we can almost promise you smooth sailing, and the undeniable thrill of watching a guest smacking his lips over a slice of your very own Gouda, and saying to him, smugly, "I made it, myself!"

It All Began . . .

To give you an idea of how old the art of cheesemaking is, we looked into the history books and discovered that man must have known about making cheese before he invented writing. A Greek historian named Xenophon, born in 349 B.C., wrote about a goat cheese that had been known for centuries in Peloponnesus.

We'll admit that knowing this hasn't improved our Cheddars at all, but it's fun to speculate on how cheese might have been discovered. We like the oft-told tale about the nomad who poured his noon ration of milk into a bottle made from a sheep's stomach and plodded across the desert all morning, only to find, at noon, that his liquid lunch had solidified!

It must have been a shock for the poor wanderer. He may not have been as brave as the man who ate the first raw oyster, but we imagine our hero eating a rather late lunch that day. Once he finally tasted it, however, he probably became the first traveling salesman, singing the praises of this new food at all his overnight oasis stops. We guess that it didn't take long for his customers to recognize a good thing when they tasted it, and that cheesemaking mystique took hold and spread across the known world.

As the center of civilization moved westward to Rome, the art of cheesemaking was carried along. The Romans refined techniques, added herbs and spices, and discovered how to make smoked cheese. Many varieties were made in those times and the Romans feasted on curd cheeses, Limburger-type, soft cheeses, smoked and salted cheeses. They exported their hard cheeses and experimented with a cheese made from a mixture of sheep's and goat's milks. They learned to use different curdling agents in addition to the rennet they extracted from the stomach of a weanling goat or sheep. Thistle flowers, safflower seeds, or fig bark were soaked in water to make extracts that would set a curd. Baskets and nets and molds were devised to shape their cheeses.

When Julius Caesar sent his soldiers out to conquer Gaul, they packed K rations of cheese for the long marches. Wherever they went and conquered and settled, the savage, northern tribesmen quickly learned to copy their captors' delicious food.

Factories, Finally

Far up in the Alps, back in the fifteenth century, Swiss farmers milked their cows out in the fields and brought the milk back to the farms to make their Emmenthaler (Swiss) cheese. It wasn't until about 1800 that they realized they could make cheese down in the valleys, as well as high in the hills. The first cheese factory was opened there in the valley, at Bern, in 1815. It was such a successful venture that in the next twenty-five years, 120 cheese factories sprouted up, with the number growing to 750 by the end of the century.

Don't paint yourself a mental picture of gleaming, stainless steel vats in bright, sterile laboratory conditions. Those early Swiss factories had a fire pit in the corner, with a copper kettle hanging over it on a crane, so they could swing the kettle over or away from the fire.

Temperature was tested on a forearm. Their rennet solution, which was a water extract of a calf's stomach lining, was so strong that they claimed it would set milk in the time it took to recite The Lord's Prayer. (This is a lovely idea, but when we tried it, reverently and slowly, we were finally able to drag it out to thirty seconds, a trifle short for setting time in most of our recipes.)

The first American cheese factory was established in Rome, New York, in 1851, by an ingenious entrepreneur named Jesse Williams. Earlier companies, as far to the west as Ohio, had been set up to buy homemade curds from local farmers and process the curds into cheeses weighing from ten to twenty-five pounds. Williams realized that cheese made from several different batches of curd would lack uniformity of taste and texture. He also knew that it takes exactly the same length of time to make a curd from 1,000 pounds of milk as it does from a pint, and he set up his factory to make cheese from scratch.

He bought milk from local farmers to add to the output of his own herd of cows and in its first years of operation, his factory produced four cheeses per day, each weighing at least 150 pounds. Spring water was circulated around the vats to cool them, and steam, produced by a wood-fired boiler, heated them. It is recorded that his costs averaged about 5½ cents a pound for an aged cheese, and that he sold his entire stock on contract for a minimum of 7½ cents a pound. The financial success of the whole operation was enhanced by the sale of pork from his large herd of hogs, who were fed the factory's whey output.

Modern scientific methods, developed throughout the latter half of the nineteenth century by men like Jesse Williams, have taken all of the guesswork out of commercial cheesemaking. Models of cleanliness and efficiency, today's cheese factories are equipped with steam-jacketed stainless steel vats, thermostatic controls, mechanical agitators, and curd cutters. Sterile conditions remove any possibility of unwanted bacterial invasion of a cheese, at any stage of its development.

Yet, we are all do-it-yourselfers at heart. In our concern for the environment, as well as for our poor, overburdened budgets, more and more of us are trying to relearn some of the techniques of self-sufficiency that made our ancestors the independent folk they were. We have learned to heat with wood, bake our own bread, and brew our own beer. Some of us have even dabbled in distilling fuel for our cars and farm tractors. From this book you can learn to make your own cheese.

EQUIPMENT YOU WILL NEED

Much of the equipment you will need for cheesemaking can be found in your kitchen. There are a few precautions that must be taken in the use of this equipment, and these are extremely important.

1. All utensils (and this includes *everything* that comes in contact with the milk at any stage) used during cheesemaking should be made of glass or stainless steel, or enamel-lined (with no chips or cracks, please). During cheesemaking, milk becomes acidic, and if aluminum or cast iron pots are used, metallic salts will be taken up into the curds, causing an unpleasant flavor and being potentially dangerous to eat.
2. All utensils must be carefully cleaned before and after cheesemaking. *Most home cheesemaking failures are caused by unclean or unsterile equipment.* Remember that the whole process of making cheese is based on the action of friendly bacteria. Unclean and unsterile conditions can add the unwanted factor of "war between the good guys and the bad guys."

Cleaning Your Equipment

The minute you're through with a utensil, rinse it thoroughly with *cold water.* Hot water will cook the milk onto the surface of a pan or ladle. Then wash and scrub it in hot water with a good dish-washing detergent or soap (or you may use washing soda). Rinse thoroughly in hot water. You must wash utensils just as carefully *before* using them.

Sterilizing Your Equipment

You may sterilize your equipment in one of three ways.

1. Immerse equipment in boiling water for up to five minutes.
2. Steam utensils for a minimum of five minutes in a large kettle with about two inches of water in the bottom and a tight lid on top. Wooden items

such as cheese boards and mats should be boiled or steamed for at least twenty minutes.

3. Plastic (even food grade) equipment should not be boiled or steamed, and should be sterilized with a solution of household bleach (sodium hypochlorite) mixed in the proportion of two tablespoons of bleach to one gallon of water. Bleach may also be used with stainless steel utensils. Be sure that your rinsing is very thorough after using bleach, because a residue of sodium hypochlorite will interfere with the growth of cheesemaking bacteria. Drain the rinsed equipment dry and store it in a clean place.

Just before using them again, all utensils should be resterilized. Dampen a clean cloth in the bleach solution and wipe all counter areas around the place where you will work.

These are some of the things you will need:

1. Glass or stainless steel *measuring cup and spoons.*
2. *Dairy thermometer.* The thermometer should measure from 20° to 220° F. Before using it, check its accuracy by testing it in boiling water. If it doesn't read exactly 212° F., you'll need to make adjustments when using it.

   There are two types of thermometers.

   One is the floating glass dairy thermometer, a fine choice for beginners. It floats upright in the milk and is easy to read.

   The stainless steel dial thermometer is faster in its response to temperature changes than the glass model. Most have a bracket to hold them at the side of the pot. They have a nine-inch shaft suspended in the milk.
3. *Pots.* They should be stainless steel or glass, or enamel-lined, and large enough to hold the amount of milk you will use.
4. *Pots.* You'll also need a pot larger than the one that contains the milk, to hold water, double-boiler fashion, around the cheese pot. This pot should have a rack, so that water can circulate under the cheese pot. As an alternative, you can put the cheese pot in a sink of hot water. In many

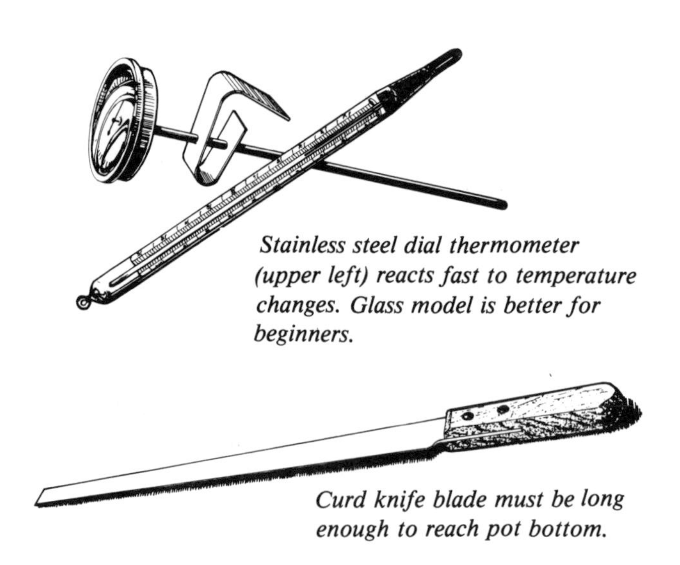

Stainless steel dial thermometer (upper left) reacts fast to temperature changes. Glass model is better for beginners.

Curd knife blade must be long enough to reach pot bottom.

cases, this is the easiest way to warm milk. When recipes in this book call for heating indirectly, use one of these methods.

5. Stainless steel stirring *spoon* or *ladle.*
6. *Curd knife* with a stainless steel blade long enough to reach to the bottom of the smaller pot without immersing the handle.
7. *Cheesecloth.* We don't mean the loosely woven mosquito netting fabric sold in most stores under the name of cheesecloth. The real thing is much more closely woven, needs only one layer of thickness for draining cheese, and is strong enough to wash, boil to sterilize, and use over and over again. After being used, cheesecloth should be rinsed clear in cold water, then washed right away with a little bleach added to the wash water occasionally. Boiling the cheesecloth in water to which washing soda has been added will help maintain the freshness of the cloth.
8. *Butter muslin.* Slightly closer weave than cheesecloth.
9. *Molds* come in many shapes and sizes and are used to contain the curds during the final draining period. When finished, the cheese retains the shape of the mold. Molds come in stainless steel and food-grade plastic. A make-do mold can be

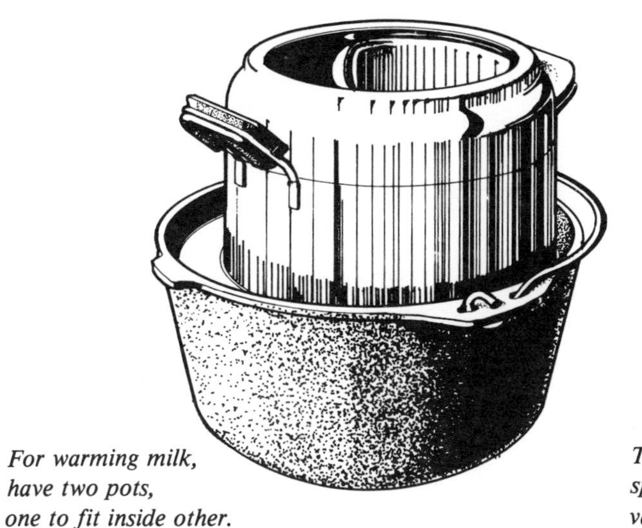

For warming milk, have two pots, one to fit inside other.

There's usually a special mold for each variety of cheese.

made by punching holes in the sides and bottom of a plastic cottage cheese container. Punch holes from the inside.

10. *Cheese press.* Essential for making a hard cheese. It should be easy to assemble, easy to clean, and have a provision for measuring the amount of pressure being applied to your cheese. Several different presses are available today, or you can build a press for yourself or wheedle one out of your next-door-neighbor's workshop — maybe in exchange for some homemade cheese. Here are some choices.

The Wheeler press is made in England from hardwood and stainless steel. A table-top model, its pressure is set by hand, and a gauge indicates pounds of pressure on the cheese — up to fifty pounds with the regular springs and up to eighty pounds with heavy-duty springs.

A homemade press can be put together from scraps of wood. Here is a design for a simple press using weights placed on the top board. Weights can be bricks, cement blocks, or a gallon jug filled with water. (A glass gallon jug of water weighs about 10 pounds.)

The Garden Way cider press can be used successfully for cheese pressing if you use a stain-

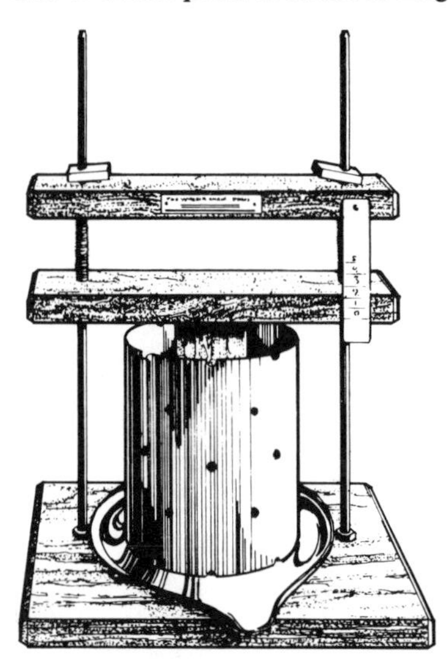

The Wheeler press, an English model ideal for beginner.

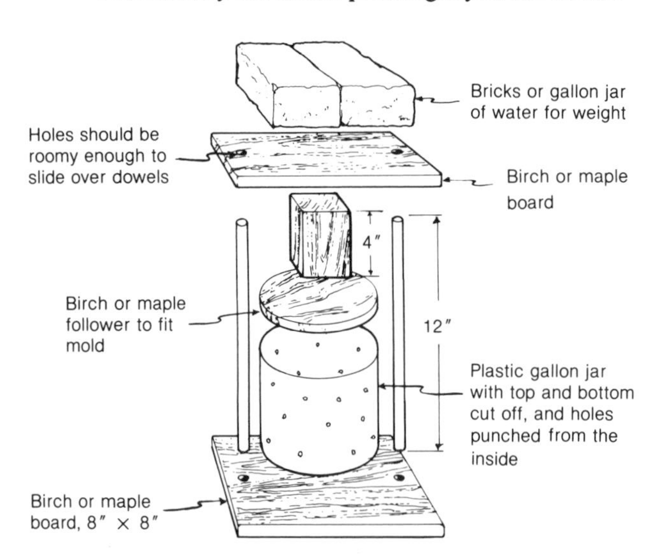

This model is easy to make, and is equally easy to use.

A mold adapts Garden Way cider press for cheesemaking.

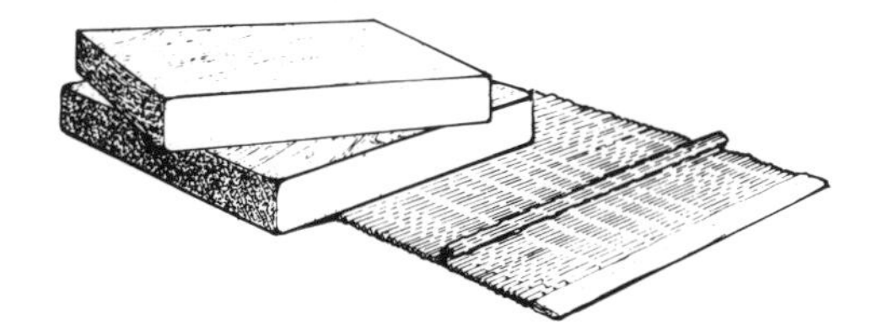

Cheese boards and mats are needed to drain cheeses.

less steel mold inside the base to contain the curds.

11. *Cheese boards* are useful as draining platforms for such cheeses as Camembert, Coulommiers, and Brie. They should be of well-seasoned hardwood and should measure about six inches square.

12. *Cheese mats* are made from wooden reeds sewn together with cotton twine, or formed in food-grade plastic. They are used to allow the whey to drain away from a cheese. Essential for Coulommiers, Brie, and Camembert, they are also useful for the air-drying period following the pressing of a cheese.

INGREDIENTS AND HOW TO USE THEM

MILK

Man has taken milk from many animals during the course of history. The familiar cow, goat, and sheep have fed people for centuries, and so have some less common animals such as the yak, camel, buffalo, llama, ass, elk, mare, caribou, and reindeer. Cow's milk and goat's milk are the only ones readily available in this country today, so they will be used in the recipes in this book.

If you know how, you can turn milk into cheese without adding a single thing. Over the centuries, man has devised hundreds of ways of doing this. Time and temperature, and the addition of a large selection of different ingredients determine the flavor and texture of each variety.

Milk is a complicated substance. About seven-eighths of it is water. The rest is made up of proteins, minerals, milk sugar (lactose), milk fat, vitamins, and trace elements. These substances are called milk solids.

When we make cheese, we cause the protein part of the milk solids, called casein, to curdle. At first, the curd is a soft solid gel because it still contains all of the water mixed in with the solids. But as it is heated, and as time passes, the curd releases the liquid, called whey, and condenses more and more — until it becomes a cheese.

Fresh milk, right from the animal, is called whole milk to distinguish it from other forms of milk that man has made for his own convenience, such as skim, homogenized, evaporated, and dried milk. Whole milk contains about 4 percent butterfat; skim milk, only 1 percent. When milk is converted to cheese, most of the fat remains in the curd, with very little going off in the whey. Homogenizing breaks up the fat globules into very small particles, and then distributes them throughout the milk, so they do not rise to the top as cream. It is more difficult to make a cheese from homogenized milk because it forms a curd less firm than one made from whole milk. Skim milk is used for making cheese starter culture, low-fat cottage cheese, and the hard, grating cheeses such as Romano and Parmesan.

One pound of hard cheese starts out as ten pounds of milk, so the non-water elements of the milk are highly concentrated when in cheese form. The following table shows how the curds and whey divide up the contents of the original milk when you make a Cheddar.

Content	*Whey*	*Curd*
Water	94%	6%
Fat	6	94
Total solids	52	48
Casein	4	96
Soluble proteins	96	4
Lactose	94	6
Calcium	38	62
Vitamin A	6	94
Thiamine	85	15
Riboflavin	74	26
Vitamin C	94	6

Goat's Milk

Goat's milk has smaller fat globules than cow's milk, and is more easily digested. Little cream rises, compared to the thick, rich layer found on a gallon of cow's milk after it has stood for a few hours, but the fat content is about the same. It is whiter than cow's milk, due to the lack of carotene, and cheese made from it is whiter than cow's milk cheeses.

These differences are important in cheesemaking, and call for different temperatures and methods, depending on which milk you are using. We'll either note the difference within each recipe, or we'll show the goat's milk recipe separately.

Milk means different things to different people. For the shopper in a grocery store milk is the white liquid found in plastic jugs in the dairy case. For the owner of a dairy animal, either a cow or a goat, milk is what you get twice a day when you do your farm chores.

There are a number of terms which need to be defined so that we all know what we are talking about when the word milk is used. These terms are: raw milk, pasteurized milk, whole milk, skim milk, and homogenized milk.

Raw milk is that which is collected directly from a dairy animal. It contains natural flora, many of which are very useful in cheesemaking. Raw milk may also contain bacteria which are harmful to man. Harmful bacteria are known as "pathogens" and they can produce diseases in man. Two pathogens which can be found in milk are tuberculosis and brucellosis. If you consume raw milk or produce cheeses with raw milk which are aged less than sixty days (this includes almost all fresh cheese) *you must be absolutely certain that there are no pathogens in the milk.* In order to be certain that your raw milk is pathogen-free you must consult a local veterinarian for advice. *A good rule to follow is: if in doubt, pasteurize.*

Also when using raw milk in cheesemaking, you must never use milk from an animal that is suffering from mastitis (an inflammation of the udder) or from an animal receiving antibiotics (for these drugs will destroy helpful bacteria whose presence is essential in the making of most cheeses).

PASTEURIZATION

Milk which has been heat treated to destroy all pathogens is known as pasteurized milk. All milk purchased in the store has been pasteurized and need not be pasteurized again. If you are acquiring milk from a dairy animal directly and need to pasteurize your milk, either cow's or goat's, here is a simple procedure to follow:

How to Pasteurize Milk

Pour raw milk into a stainless steel pot and place this pot in another, larger pot containing hot water. Put the double boiler on the burner and bring the water to a boil.

To pasteurize, hold milk at 145 ° F. for 30 minutes.

Then cool milk as rapidly as possible in ice water.

Heat the milk to 145° F., stirring occasionally to insure even heating. Hold the temperature at 145° F. for 30 minutes exactly. The temperature and time are important. Too little heat and too short a holding time may not destroy all the pathogens. Over-pasteurization can result in a curd too soft for cheesemaking due to the destruction of the milk protein.

Remove the milk from the hot water and put it in a sink filled with ice water to the same level as the milk. Stir constantly until the temperature drops to 40° F. Rapid cooling is important. Store pasteurized milk in the refrigerator until you're ready to use it.

Whole milk contains cream. Skim milk has had

almost all of the cream removed. Homogenized milk has had the fat globules of cream mechanically broken up into such small sizes that the cream will remain dispersed in the milk and will not rise to the top as it will in non-homogenized milk. If possible it is best to use non-homogenized milk in cheesemaking.

Using Milk Powder

You can make cheese with whole milk, raw milk (when cheese is aged more than sixty days), skim milk, pasteurized milk, or even with reconstituted dry milk powder. One and one-third cups of dry milk powder dissolved in three and three-quarters cups of water make one quart of milk. It doesn't need to be pasteurized because the drying process destroys the unwanted bacteria.

The milk you use for cheesemaking must be of the highest quality. Buy the freshest milk possible. If it comes from the supermarket, don't open the container until you're ready to start, so it won't be contaminated with unwanted bacteria from the air. Above all, don't make cheese with milk that tastes "off."

Coagulation — How and Why

The first step in cheesemaking is to coagulate the milk solids into a curd. There are two basic ways to do this. Each way has many variations; used together in combination, the variations are endless. The two methods are *acid coagulation* and *rennet coagulation.*

You can cause acid coagulation either by adding an acid substance such as lemon juice or vinegar, or by adding a bacterial culture which turns the lactose (milk sugar) into lactic acid. Both are good ways to make a curd, and each is used for different purposes.

To see the first method work, try making a small batch of lemon cheese. Heat one pint of milk to 100° F. in a double boiler over simmering water. Take the milk pan out of the hot water and add the juice of one lemon. Stir it well and leave it for about fifteen minutes. As it sits, you can watch a rather stringy curd

In making lemon cheese, add juice to milk at 100° F.

appear and see the white milk liquid turn into a milky whey.

Drape a piece of sterilized cheesecloth over a colander set inside a larger bowl. Gently ladle the curds into the cheesecloth with a slotted spoon or ladle. Hold three corners of the cheesecloth in one hand and wrap the fourth corner around the other three, poking the end down through the loop it makes. Hang the bag of curds over a bowl and drain it for about an hour. Your lemon cheese will then be ready to eat. It will be moist and spreadable and have a milky, slightly lemony flavor. Store it in the refrigerator. Spread it on whole wheat bread for a delicious snack.

Ladle curds into cheesecloth-covered colander in bowl.

Bacterial Culture

The second method of acid coagulation begins with adding a bacterial cheese starter culture to milk at room temperature (72° F.). Cheese starter culture bacteria are one-celled living organisms who live by eating milk sugar (lactose). When added to milk at room temperature these bacteria will consume lactose and produce a byproduct, lactic acid. As time passes these bacteria reproduce rapidly and after fifteen to twenty-four hours there are an astronomical number of starter bacteria living in the milk.

They have produced so much lactic acid that the milk protein coagulates into a solid white gel known as the curd. This curd can be used in cheesemaking for cheeses such as cottage cheese or lactic cheese. This method of coagulation actually places acid makers (starter bacteria) into the milk and over a period of time they produce sufficient lactic acid to coagulate the milk.

Bacterial cultures are a very important part of cheesemaking. It is best to prepare your starter prior to making any cheese for which it is needed.

Cheese Starter Culture

Cheese starter culture is a growth of specific bacteria in sterile milk. It assures that the right proportion of lactic acid-producing bacteria will be growing in the milk you use to make cheese. These bacteria change the milk sugar (lactose) into lactic acid, which is the basis of the long-keeping quality of cheese.

It is important to develop the proper amount of acidity in the milk. The acidity which the cheese starter culture bacteria produce helps the rennet to coagulate the milk, aids in expelling the whey from the curds, and checks the growth of pathogens in the finished cheese. The starter culture bacteria are also responsible for much of the flavor development in an aging cheese.

Milk that is increasing in acidity due to starter activity is referred to as ripening. Milk that has reached the proper degree of acidity is referred to as ripened.

The two basic categories of cheese starter culture used in cheesemaking are mesophilic and thermophilic cultures. A mesophilic (moderate temperature loving) culture is used in cheese where the curds are not warmed to over 102° F. during cooking. This would include cheeses such as Cheddar and Gouda.

A thermophilic (heat-loving) culture is used in cheeses in which the curd is cooked at temperatures up to 132° F. The bacteria thrive at high temperatures. Swiss cheese and the Italian cheeses need such a culture.

There is a large variety of starter cultures for producing an enormous selection of cheese, but all are either mesophilic or thermophilic cultures.

1. Cheese starter culture (mesophilic) produces many different hard cheeses including Cheddar, Gouda, farmhouse, and Caerphilly.
2. Cheese starter culture (thermophilic) produces Swiss and Italian-type cheeses.
3. Sour cream, buttermilk, and fresh starter culture (mesophilic) produces a variety of soft cheeses and contains extra flavor-producing bacteria.
4. Goat cheese starter culture (mesophilic) produces excellent goat cheeses and is desirable because it contains a minimum of flavor-producing bacteria.

Starter cultures are sold as a freeze-dried powder, and come in a foil packet which must be stored in the freezer until used. Starter culture must be prepared before it can be used in cheesemaking. When you are ready to make the mother culture, you should follow these steps very carefully. (Mother culture means that you can use your first batch to make more later on.)

How to Prepare Start Culture (Mesophilic)

1. Sterilize a clean one-quart canning jar and its cover (or two pint-size canning jars with covers) by placing them in boiling water for five minutes.
2. Cool them and fill the jar with fresh skim milk, leaving 1/2 inch of head space. Cover the jar tightly with its sterilized lid.
3. Put the jar in your canner (or a big, deep pot) with the water level at least 1/4 inch over the top of the jar lid.
4. Put the pot on the burner and bring the water to a boil. Note when the water begins to boil, and let it continue at a slow boil for thirty minutes.
5. Take the jar out of the water, and let it cool to 72°, away from drafts. (To check temperature,

To prepare starter culture, sterilize milk in hot water.

Pour culture into sterilized milk cooled to 72° F.

use the room temperature, and avoid contaminating the milk with your thermometer.)

6. Inoculate the milk by adding the contents of the freeze-dried starter culture packet to the milk (still at 72°). Add the powder quickly, to minimize exposure to the air. Re-cover and swirl the jar every five minutes or so, to mix and dissolve the powdered culture thoroughly. (If you're making your second batch, you'll be adding two ounces of fresh or frozen starter to the sterilized milk.)
7. Put the jar where the milk temperature can be kept at 72° for fifteen to twenty-four hours during its ripening period. Sixteen hours usually does the trick, but if the milk hasn't coagulated by then, you can leave it another eight hours, or a little more.
8. The culture will have the consistency of a good yogurt. It will separate cleanly from the sides of the jar, and its surface will be shiny. Taste it. It should be slightly acid and also a bit sweet.
9. Chill it immediately. You can keep it in the refrigerator for up to three days before using it. Unless you plan to make a large amount of cheese right away; however, the best thing to do is to *freeze it for storage.*
10. To freeze starter culture, sterilize four plastic ice cube trays. Fill the trays with the starter culture

To store culture, freeze it in sterilized ice cube tray.

and freeze it solid in the coldest part of your freezer. Transfer the cubes to airtight plastic bags and put them back in the freezer. They'll keep their strength for up to one month. Each cube of starter culture is a convenient one ounce, which can be thawed at any time and used to make a cheese, or another batch of starter culture.

Preparing Thermophilic Starter Culture

Follow the directions above with these exceptions:

In step 6, inoculate the milk with thermophilic culture when the temperature is 110° F.

In step 7, allow the milk to incubate at 110° F. for six to eight hours until a yogurt-like curd is produced.

What if Something Goes Wrong?

If you've been careful about sterilizing everything, and about the timing and the temperature rules above, probably nothing will go wrong. But . . .

1. If your starter tastes sharply acid, or even slightly metallic, it may mean that it has overripened. (Next time, use a little less starter, or incubate it at 70° instead of 72°.)

2. If your starter won't coagulate, it could mean any (or all) of the following:

a. The temperature dropped below 72° (110° F. for thermophilic culture) during the ripening period.

b. The inoculating culture didn't contain live bacteria.

c. The milk contained antibiotics. This happens occasionally when a dairy must give an antibiotic to a cow. The medicine is absorbed in the animal's system and comes out in the milk.

d. In cleaning your utensils, you used a bleach or a strong detergent and didn't rinse thoroughly enough. Residual amounts of either can halt bacterial action.

e. You didn't add enough starter culture. This is unlikely in your first batch, because the packets of freeze-dried culture are very carefully premeasured. In your second and succeeding batches, be sure to add two ounces of fresh starter, or two cubes (two ounces) of frozen starter culture, from your freezer supply.

f. Also unlikely, but still possible, is that organisms hostile to the lactic acid-producing bacteria are present in the culture.

3. If you find bubbles in your finished starter culture, it could mean:

a. The skim milk was not properly sterilized in the canning jar step.

b. Your equipment was not clean enough.

Bubbles in starter are manufactured by gas-producing organisms such as yeasts or coliform bacteria. They are present in a starter due to faulty preparation technique, and such a starter should be discarded.

If you have *any* reason to believe that your starter culture isn't quite right, *throw it away and begin again with a fresh culture.* It would be heartbreaking to wait six long months for your cheese to age, only to find that the wrong bacteria have been at work spoiling it.

Buttermilk

Originally buttermilk was what was drained from the churn after butter had been made.

Little of that is available today. Instead, the buttermilk you buy is made from pasteurized skim milk to which a culture of bacilli has been added.

In the recipes calling for cultured buttermilk this is what you will use. You can make your own cultured buttermilk.

Those familiar with making yogurt will recognize the two methods that can be used.

One is to buy the freeze-dried buttermilk culture and add that to the skim milk.

The other is to buy fresh cultured buttermilk and use that as follows:

1 quart skim milk

½ cup cultured buttermilk.

In either case, heat milk to 70° F., add culture or buttermilk, then stir well. Cover and let stand at room temperature until the milk has clabbered (with the appearance of sour milk). Stir until smooth, then refrigerate.

Rennet

Cheese rennet is not the same as the junket rennet sold in grocery stores. It is possible to curdle milk with junket rennet, but the resulting curd, while it makes a pleasant dessert, will not make an acceptable cheese. Cheese rennet is available as tablets or in liquid form.

Rennet that is an animal derivative is extracted from the fourth stomach of a calf or young goat. Its rennet contains an enzyme called rennin which has the property of causing milk to form a solid curd. In the days before modern laboratory technology, which can now produce a standardized rennet, most cheesemakers made their own on the farm. When they slaughtered a calf or kid, they cleaned and salted the stomach and hung it up to dry. Then it was stored in a cool place until they needed it. At cheesemaking time, they broke off a small piece of the dried stomach and soaked it in cool, fresh water for several hours, then added a bit of the solution to the ripened milk to produce a curd.

Another down-on-the-farm method of making rennet involved the stomach of a calf or a kid slaughtered at not more than two days old. The stomach would contain milk with a high percentage of colostrum. Taking this milk out and cleaning the stomach carefully, inside and out, the dairyman (or maid) would then return the colostrum-rich milk to the stomach, tie off the tip, and hang it in a cool place to age. When cured, the contents would have turned into a lard-like substance, which would then be kept as cold as possible, stored in a tightly covered container. A "thumbnail's worth" of this paste would set at least two gallons of milk. With this method, it was also possible to add a tiny amount of finely grated, dried cheese to the milk as it was replaced in the stomach, producing a finished rennet that was culture and coagulant all in one.

There are a number of plants that have coagulating properties. In ancient Rome, cheesemakers used an extract of fig tree bark. An infusion made of the weed called Lady's Bedstraw or (*Galium verum*) the stinging nettle (*Urtica dioica*) can be used. The flower of the thistle plant called *Cynara cardunaculus* is used in Portugal to make Sera de Estrella cheese.

It is said that at one time, in northern Europe, the plant called butterwort was fed to the cows just before milking time, and that it caused the milk to coagulate, three hours after milking time.

Today, rennet is available as both animal and vegetable derivatives. Vegetable rennet is an enzyme derived from the mold *Mucor miehei,* and is marketed in liquid and tablet form. Commercially available rennet is standardized today in the United States to a strength of 1 to 15,000 (in Europe to 1 to 10,000). This means that one part of rennet will coagulate 15,000 (or 10,000) parts of milk, depending upon where you buy your rennet. If you are using a modern European recipe, compensate slightly for the difference in rennet strength.

You can buy animal rennet in tablets or in liquid form. Tablets are easier to store, and keep longer than the liquid. Liquid rennet is easier to measure accurately. It must be refrigerated and kept away from long exposure to light.

Rennet Coagulation

When milk has ripened for the proper length of time for the cheese you plan to make, and it is still at the right ripening temperature (85° to 90° F.), it is time to add rennet. Before rennet is added to the ripened milk, it must be diluted in sterile water (water that has been boiled and then cooled to room temperature). Liquid rennet should be diluted in about twenty times its own volume of water. A rennet tablet (or fraction of one) is crushed, then dissolved in about 1/4 cup of water. The correct amounts of rennet to use will be found in each recipe.

Sprinkle the diluted rennet over the surface of the milk and stir it in thoroughly — all the way to the bottom of the pot — to insure an even set. Then cover the pot and let it sit undisturbed for about forty-five minutes at the same temperature. A curd is set when a finger (or a dairy thermometer) is inserted into it and it breaks cleanly all around as you lift slightly.

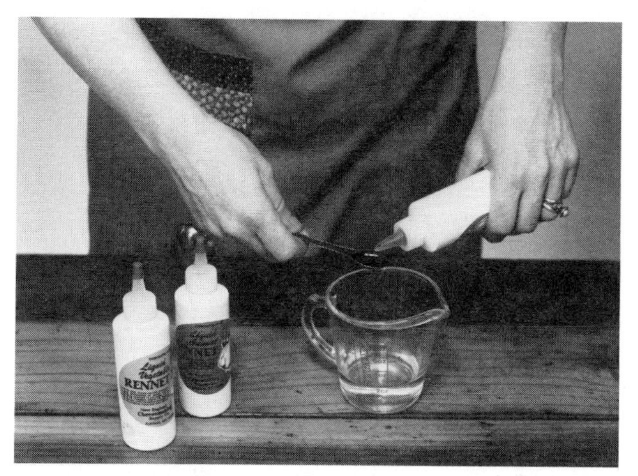

To dilute liquid rennet, add twenty times as much water.

Temperature

Coagulation is a complicated phenomenon. Time, temperature, acidity, and the amount of rennet used all play a part in its functioning. Rennet usually works most efficiently at 104° F. If the milk is warmer or cooler than that, the action slows down. Rennet also works faster in milk with a higher acid content. We can hear you asking, "Then why do we add rennet at 85°F, when it works better at 104° F?" The answer is, "To give us control of coagulation."

Curd starts out as a soft mass, and increases in firmness as time passes, until it reaches the point when it's best for cheesemaking. Beyond that point, the quality deteriorates. We lengthen the time element by controlling acidity, temperature, and the amount of rennet, so we'll be able to identify the exact moment when the curd is ready to cut. If we start too soon, the curd will be too soft to be workable, and if we wait too long, it will become weak.

We control acidity by planning the ripening time so that the level of lactic acid will be just right. Too little acidity makes a weak curd, and too much may produce a sour, bitter-tasting cheese. Experienced cheesemakers use acid-testing equipment to monitor development of lactic acid at all stages of cheesemaking.

Measure the right amount of rennet carefully. Too little doesn't set the milk at all, and too much makes cheese taste bitter. The perfect curd is the one that produces the highest yield of cheese from any given amount of milk.

It is important to follow recipe directions carefully. Each recipe has been tested for correct times, temperatures, and amounts, to give you the good results you want. Because there are so many variables, it is also important to keep careful records as you go along. This will help you to repeat your successes exactly, or to correct a mistake, should you have the misfortune to make one. (To help you take the best kinds of notes, you will find a sample cheese record form on page 23.)

If the cheese doesn't come out exactly as it should, try looking through the list of possible problems, in the trouble-shooting chart on page 125.

Coloring

The characteristic color of a finished cheese is as much a part of its identity as its flavor. Goat's milk produces a very white cheese, due to the lack of carotene in the milk. At the other end of the color spectrum is the deep orange of the longhorn and Colby cheeses. Cow's milk, with a much higher carotene content, makes a perfect Cheddar color with no additive at all. Color has no effect on the flavor of a cheese except for Roquefort and blue-type cheeses in which the identifying blue flecks are caused by the same mold or fungus which imparts their delectable flavor.

In times past a yellow cheese was a higher quality cheese since it was made from a milk rich in butterfat. Cheesemakers soon discovered that they could artificially color their cheeses and command a premium price for cheeses made with inferior milk. At one time marigold petals were used for coloring. Saffron, tumeric, and hawthorne buds were also used.

In the eighteenth century a derivative of the annatto tree was imported from the West Indies to color cheeses and to this day annatto is the favored ingredient. Annatto comes in both tablet and liquid forms and is a safe, non-toxic vegetable dye. It is diluted in water and stirred into the milk after the starter culture has ripened the milk and before the rennet is added. Wash the container carefully if you will use it for rennet, because the annatto, if concentrated, will weaken the action of the rennet. It should be used very sparingly. It will hardly show in the milk when you stir it in, and becomes more pronounced as the curd condenses.

Cheese coloring becomes more pronounced as curds form. A coarse flake salt is recommended for all cheesemaking.

CHEESE RECORD FORM

Type Of Cheese ____________________ Date ________ Type of Milk ____________________ Amount Of Milk ____________

1. **RIPENING:**

   Type of Starter ______________________________
   Amount of Starter ______________________________
   Time at Adding Starter______________________________
   Milk Temperature at Time of Adding Starter ______________

2. **RENNETING:**

   Type of Rennet ______________________________
   Amount of Rennet ______________________________
   Time at Adding Rennet ______________________________
   Milk Temperature at Time of Adding Rennet ______________

3. **CUTTING THE CURD:**

   Size of the Curds ______________________________
   Time at Cutting Curds ______________________________

4. **COOKING THE CURD:**

   Time of Cooking Curd ______________________________
   Temperature at Start of Cooking ______________________
   Temperature at finish of Cooking ______________________

5. **DRAINING THE CURD:**

   Time of Draining ______________________________

6. **MILLING THE CURD:**

   Time of Milling ______________________________

7. **SALTING THE CURD:**

   Amount of Salt Added ______________________________
   Type of Herbs Added ______________________________
   Amount of Herbs Added ______________________________

8. **PRESSING THE CURD:**

   Time at Start of Pressing ______________________________
   Amount of Pressure at Start ______________________________
   Amount of Pressure at End of Pressing ______________
   Date at End of Pressing ______________________________

9. **AIR DRYING:**

   Date Started ______________________________
   Date Finished ______________________________

10. **WAXING:**

    Date Waxed ______________________________

11. **AGING:**

    Temperature During Aging ______________________________

12. **EATING:**

    Date of First Bite ______________________________

COMMENTS AND OBSERVATIONS:

Salt

A coarse flake salt, similar to pickling salt, should be used. The Diamond Crystal salt company sells a crystal kosher salt which is very good. The salt is usually added to the curds just before they are pressed, and in some cases is rubbed gently on the outside of the cheese after the skin (or rind) has formed.

Cheese Wax

After the air-drying period, when a hard cheese has developed a dry rind, it must be protected with wax for the long aging period. You may melt paraffin (in a double-boiler is the only safe way) and add an equal amount of a good grade of vegetable oil, painting this mixture on your cheese with a bristle (not nylon) brush. It may take two or three thin coats of this mixture to protect the cheese.

Real cheese wax is stronger and more pliable than paraffin. It can be removed in two or three pieces from a cheese that is ready to be eaten. It can be melted and used again. It is applied with a bristle brush at 240° F., and kills any surface bacteria and flashes off any moisture on the cheese. Being more pliable, it will not crack as the curing cheese is turned from time to time and therefore admits no unwanted mold during aging.

Herbs

Herbs will add a variety of flavors to your soft cheeses. They also add a touch of color. It is preferable to use fresh herbs whenever possible. Some of the herbs which we enjoy using from our garden are:

Chives	*Parsley*	*Thyme*	*Garlic*
Dill leaves	*Oregano*	*Basil leaves*	*Sage*

These are best picked fresh and used in cheese as soon as possible. Amounts to use depend on your own taste buds. Soft cheeses with herbs added should be allowed to set in the refrigerator for a day or two in order for the flavors to permeate the cheese.

When using dry herbs, allow the cheese to remain refrigerated for several days, since the herbs are not as aromatic as the fresh ones. There are dry herb mixtures available specifically for making soft cheeses. Sources for these can be found in the appendix.

Of all the cheeses to use for a soft herb cheese, the lactic cheese is the all-around best choice. It readily lends itself to produce a variety of delicious cheese spreads.

Before You Begin

Just as we must all walk before we can run, it's smart to start cheesemaking with the simpler varieties.

That's why this book is set up in the order you'll find on the rest of its pages. Begin with the basic soft cheese recipe and make a few soft cheeses before you graduate to the complexities of the hard cheese section. We mentioned in the introduction that "there's a trick to it." By the time you reach the basic hard cheese recipe, you'll have mastered the art of cutting and cooking a curd, and you'll be ready to go on—successfully.

The cheeses in this book have been divided into six sections: soft cheeses, hard cheeses, whey cheeses, goat cheeses, soft bacterial and mold-ripened cheese, and cheese spreads.

Each group was assembled with the needs of the cheesemaker as the primary criterion. Thus each chapter includes a variety of cheeses which are all made in a similar fashion and involve the same basic steps and the same equipment.

There are a large number of cheeses included in this work. For the beginning cheesemaker we would suggest starting with a small number of recipes until success is achieved.

In all attempts, keep accurate records so that you can profit from your successes and learn from your mistakes.

The cheeses we recommend for beginners to try are:

Queso Blanco	**Lactic Cheese**	**Buttermilk** *(moist)*
Yogurt	**Kefir**	**Lemon**
Cottage *(large curd)*	**Cottage** *(small curd)*	**Ricotta**
Mysost	**Gjetost**	**Feta**

SOFT CHEESES

The cheeses included in this chapter are soft cheeses. They require little equipment to produce, are eaten fresh, and are excellent cheeses for the beginning cheesemaker to attempt.

Soft cheeses are high-moisture cheeses. They must be consumed fresh and will keep under refrigeration for several weeks at the most. These are for the most part non-pressed cheeses. They are divided into two categories: the bag cheeses and the cottage cheeses. For those recipes in which vinegar is called for in coagulating the milk, either cider vinegar or wine vinegar may be used. Where the recipe calls for warming the milk, do this *indirectly* with the cheese pot resting in a bowl or sink of warm water, unless the recipe states that you should heat the milk over a flame, as in Queso Blanco.

SOFT CHEESES

Variety	*Amt. Milk*	*Type of Milk*	*Coagulation*	*Culture*	*Amt. of Cheese*	*Use of Cheese*
Queso Blanco	1 gallon	Cow's, Goat's	1/4 cup vinegar	None	1 1/2–2 lbs.	Cooking
Lactic	1 gallon	Cow's, Goat's	Lactic acid rennet, if goat's milk	Mesophilic	1+ pound	Spread, dessert, cooking
Buttermilk, moist	1 qt. buttermilk	Cow's, Goat's	1 tsp. diluted lactic acid	Buttermilk (mesophilic)	6–8 oz.	Spread
Buttermilk, dry	1 qt. buttermilk	Cow's, Goat's	lactic acid	Buttermilk (mesophilic)	6–8 oz.	Spread

Variety	*Amt. Milk*	*Type of Milk*	*Coagulation*	*Culture*	*Amt. of Cheese*	*Use of Cheese*
Real buttermilk	1 gal. buttermilk	Buttermilk from Buttermaking	lactic acid + heat	None	4–6 oz.	Spread
Yogurt	1 qt. yogurt	Cow's, Goat's	yogurt (thermophilic)	yogurt	6–8 oz.	Spread
Kefir cheese	1 qt. kefir	Cow's, Goat's	kefir (lactic acid)	kefir	6–8 oz.	Spread
Lemon	1 quart	Cow's, Goat's	2 lemons	None	6–8 oz.	Spread, cooking
Neufchatel	1 gallon 1 pt. heavy cream	Cow's	Rennet 1 tsp. diluted	Mesophilic 4 oz.	1+ lb.	Spread
Gervais	1 1/3 cups cream 2 2/3 cups milk	Cow's, Goat's	Rennet 1 drop	Mesophilic 2 oz.	4–6 oz.	Spread
Bondon	1 quart	Cow's, Goat's	Rennet 1 drop	Mesophilic 2 oz.	4–6 oz.	Spread
Cream cheese uncooked curd	1/2 gal. cream	Cow's	lactic acid	Mesophilic 4 oz.	6–8 oz.	Spread, cooking
Cream cheese cooked curd	1/2 gal. cream	Cow's	Rennet + lactic acid 1 tsp.	Mesophilic 4 oz.	6–8 oz.	Spread, cooking
Swiss cream cheese	1 qt. cream	Cow's	Rennet + lactic acid 1 drop	Mesophilic 2 oz.	4–6 oz.	Spread, cooking
French cream cheese	2 cups milk, 2 cups cream	Cow's	Rennet + lactic acid 1 drop	Mesophilic 1 oz.	4–6 oz.	Spread, cooking
Cottage, small curd	1 gal. skim	Cow's	lactic acid	Mesophilic 4 oz.	1 1/2 lbs.+	Table, cooking
Cottage, large curd	1 gal. skim	Cow's	Rennet 1/4 tab., 1/4 tsp.	Mesophilic 4 oz.	1 1/2 lbs.+	Table, cooking
Cottage, Goat's milk	1 gal.	Goat's	Rennet + lactic acid 1 drop	Mesophilic 4 oz.	1 1/2 lbs.+	Table, cooking

BAG CHEESES

Many of the soft cheeses in this section are often referred to as "bag cheeses." They are made by coagulating milk or cream with cheese starter culture or an acid such as vinegar. Some recipes call for the addition of a small amount of rennet. The resulting curds are then drained in a "bag" of cheesecloth. Most of these cheeses have the consistency of a soft cheese spread, have a high moisture content, and will keep up to two weeks under refrigeration. These are delicious cheeses which can be varied considerably with the addition of herbs. For the beginner, these cheeses are ideal. They are easy to make, and involve little equipment.

These cheeses should be drained in a kitchen where the temperature stays at 72° F. Too high a temperature, as in the dog days of August, and you will have problems with yeast which can produce a gassy, off-flavored cheese. Too low a temperature and the cheese will not drain properly.

Queso Blanco

Queso Blanco is a Latin American cheese. The name means white cheese. There are many variations of this cheese throughout Latin America. It is hard and rubbery, with a bland, sweet flavor. It is excellent for cooking, and has the unique property of not melting even if deep-fried. A similar cheese, Panir, is made in India, and there are many delicious Indian recipes using this cheese.

Queso Blanco is an example of acid precipitation of milk protein. Both the casein and the albuminous protein are precipitated out into this cheese. It is easy to make, and is an excellent choice if you are in a hurry or if the weather is very hot, a condition which causes problems in the making of many cheeses.

Queso Blanco is often diced into half-inch cubes. That way it can be used in many ways, stir-fried with vegetables, added to soups or sauces (such as spaghetti), or used in Chinese cooking, as a substitute for bean curd. It browns nicely and takes on the flavor of the food and spices in the recipe. *Makes 1 1/2–2 pounds.*

ACID COAGULATION

180° F. **1 gallon whole milk**

Over a direct source of heat warm 1 gallon of milk to 180° F., stirring often to keep it from scorching. Maintain this temperature for several minutes.

10 minutes **180° F.** **¼ cup vinegar**

Slowly add vinegar until the curds separate from the whey. Usually ¼ cup of vinegar will precipitate 1 gallon of milk.

DRAINING

3 hours 10 minutes — **Cheesecloth** — **Colander**

Pour the curds and whey into a cheesecloth-lined colander. Tie the four corners of the cheesecloth into a knot and hang to drain for several hours or until the bag of curds stops dripping.

STORING

Take the mass of curds out of the cheesecloth. It will be a solid bag of curd. It may be wrapped in Saran Wrap and stored in the refrigerator until needed. It will keep up to 1 week.

Lactic Cheese

This is a delicious, soft, spreadable cheese which keeps for up to 2 weeks under refrigeration. Either cow's or goat's milk may be used. You may add herbs to it in any number of combinations for truly tantalizing taste treats. You can roll it up in a crepe with a fruit sauce for a gourmet dessert. It is easy to make and is ready to eat in 24 to 72 hours. *Makes nearly 2 pounds.*

RIPENING (RENNETING AND COAGULATION)

24 hours

72° F.
4 ounces mesophilic cheese starter culture
1 gallon whole (or skim) milk
1 teaspoon diluted rennet

To 1 gallon of whole milk (or 1 gallon skim milk if you wish to make a low-fat but drier cheese) at 72° F., add 4 ounces of mesophilic cheese starter culture and stir in thoroughly. Add 1 teaspoon of diluted rennet. To prepare the diluted rennet, place 3 drops of liquid rennet in 1/3 cup of cool boiled water and stir. Cover the container and leave undisturbed at 72° F. for 12 to 48 hours or until a solid curd forms.

First step in making lactic cheese is to heat one gallon of whole milk (or skim milk) to 72° F.

Next step is to add four ounces of mesophilic cheese starter culture, stir, cover, and wait for curd to form.

DRAINING

1 to 3 days When you uncover the pot, you will find a solid curd that looks like yogurt. Line a colander with cheesecloth and pour the pot of curds slowly into it. Knot the four corners of the cloth and hang the bag to drain for 24 to 72 hours, depending on the consistency you wish your cheese to attain. Draining 12 to 24 hours produces a cheese dip consistency. The kitchen temperature should be 72° F. to get proper drainage. If you wish the cheese to drain more quickly, you may change the cheese occasionally into a fresh cheesecloth.

MIXING, SALTING, AND SPICING

Salt **Herbs**

Put the curds in a bowl, salt to taste, and mix in herbs if you wish. A tasty cheese can be produced by adding fresh-ground black pepper, a chopped clove of garlic, chopped chives, and a dash of paprika. You will have made a little less than 2 pounds of cheese. Store it in a covered dish in the refrigerator until eaten. (If using goat's milk and the cheese has a hard, rubbery texture, try adding less rennet the next time. If the cheese is too moist, try adding a little bit more rennet.)

Next, pour curds into cheesecloth-lined colander.

Solid mass of curds look like this when whey is removed.

When draining, let whey drip into bowl placed underneath.

Food processor can be used when adding spices to cheese.

Buttermilk Cheese

Moist

This cheese is made from a cultured buttermilk. It is best to use fresh, homemade, cultured buttermilk made from a packet of buttermilk starter, as this will have a thick, clabbered consistency. Storebought buttermilk is thinner, and, if used, should be poured through 2 layers of cheesecloth. To thicken the thinner buttermilk, add one drop of liquid rennet to six teaspoons of cool water, then add 1 teaspoon of this to the buttermilk. Let it set for several hours at 72° F.

This cheese is fairly moist because the curds are not heated before draining. *Makes 6–8 ounces.*

DRAINING

12–24 hours — **72° F.** — **1 quart freshly made cultured buttermilk** — **Cheesecloth** — **Colander**

Pour a quart of fresh buttermilk at room temperature into a cheesecloth-lined colander. Tie the four corners of the cheesecloth and hang this bag to drain for 12 to 24 hours, or until the bag stops dripping.

MIXING, SALTING, AND SPICING

Salt — **Herbs**

Place the cheese in a bowl. Add a pinch of salt or herbs if you wish. Store in the refrigerator.

Buttermilk Cheese

Dry

This cheese is made from fresh, homemade cultured buttermilk. Because it is heated, this cheese has a fairly dry texture. *Makes 6–8 ounces.*

COOKING

160° F. — **1 quart fresh cultured buttermilk**

Heat fresh cultured buttermilk, preferably homemade, indirectly to 160° F., stirring now and then. The curds will separate from the whey.

DRAINING

6 to 12 hours — **Cheesecloth** — **Colander**

Pour the curds into a cheesecloth-lined colander, tie the four corners of the cheesecloth, and hang the bag to drain for 6 to 12 hours or until the curds have stopped dripping.

MIXING, SALTING, AND SPICING

Salt — **Herbs**

Place the curds in a bowl. If desired, add

salt and herbs to taste. This cheese will have a grainy, spreadable texture and a slightly acid taste.

Real Buttermilk Cheese

This cheese is made from "real" buttermilk. When cream is churned, butter is produced, and the liquid left in the butter churn is called buttermilk. This buttermilk can be used in cooking (buttermilk pancakes) or to make a cheese.

The cheese has a spreadable texture and a light, sour flavor. It cannot be purchased in any cheese store. *Makes 4–6 ounces.*

RIPENING

24 hours **72° F.** **1 gallon fresh buttermilk**

Allow 1 gallon of fresh buttermilk to set at 72° F. for 24 hours. This will lightly sour the buttermilk. If you desire a less sour cheese, omit this step and use fresh buttermilk.

COAGULATION AND COOKING

160° F.

Heat the buttermilk to 160° F. At this temperature the buttermilk will separate into curds and whey.

DRAINING

27-28 hours **Cheesecloth** **Colander**

Pour the coagulated buttermilk into a cheesecloth-lined colander. Use a butter muslin quality cloth (a very fine weave). Tie the four corners of the cheesecloth into a knot and hang the bag to drain for 3 to 4 hours.

MIXING AND SALTING

Salt

Place the cheese in a covered container. Salt may be added if desired. Store in the refrigerator.

Yogurt Cheese

This cheese has the tart flavor characteristic of a fresh yogurt. It is a tasty treat on crackers and can be enhanced with the addition of herbs. It is best made from fresh yogurt, however store-bought yogurt also works well. *Makes 6–8 ounces.*

DRAINING

12 to 24 hours **72° F.** **1 quart yogurt** **Colander** **Cheesecloth (fine weave butter muslin)**

Pour fresh yogurt at room temperature (72° F.) into a cheesecloth-lined colander. The cheesecloth should be of a very fine weave or all will drain out and be lost. Tie the four corners of the cheesecloth and hang to drain 12 to 24 hours or until the bag of yogurt has stopped draining.

MIXING, SALTING, AND SPICING

Salt **Herbs**

Take the cheese from the bag of cheesecloth. You may have to scrape some off the sides. You can add salt or herbs to taste if you like. Store in the refrigerator until consumed.

You can produce a tasty dessert cheese by adding the following to the yogurt cheese and mixing well:

1 teaspoon cinnamon
2 tablespoons honey
¼ teaspoon allspice
¼ teaspoon nutmeg
¼ teaspoon vanilla

For a flavorful herb cheese add and mix well:

2 tablespoons chopped fresh chives
1 teaspoon chopped fresh dill leaves
1 small clove garlic, chopped
¼ teaspoon fresh ground black pepper
Salt to taste (my taste is a pinch)

Kefir Cheese

Kefir is a cultured milk drink which has been used for thousands of years in Caucasia, a region in Russia. It is made by adding kefir culture to fresh whole milk, which is coagulated in a curd by the culture and fermented slightly with a yeast. It has a light, bubbly sparkle and is often referred to as the "champagne of milk." This is a delicious summertime drink.

Kefir can be made into a very tasty kefir cheese. *Makes 6–8 ounces.*

DRAINING

12–24 hours

72° F.
1 quart kefir
Colander
Cheesecloth (fine weave butter muslin)

Use 1 quart of fresh kefir at room temperature (72° F.). A homemade kefir is preferable. It will have a much thicker curd than commercial kefir. Pour the kefir into a cheesecloth-lined colander and tie the four corners of the cheesecloth into a knot. Hang the bag to drain for 12 to 24 hours or until the bag has stopped draining.

MIXING, SALTING AND SPICING

Salt **Herbs**

Place the drained cheese into a bowl.

You may have to scrape the cheesecloth a bit. If desired, salt and herbs may be added.

Lemon Cheese

This cheese has a delicate flavor of lemon. It is a moist cheese with a spreadable texture. It can be used as a spread or in cooking. *Makes 6–8 ounces.*

ACIDIFYING AND COAGULATION

15 minutes	**170° F.** **Juice of 2 lemons**	**1 quart milk**

Indirectly heat 1 quart of milk to 170° F. Add the juice from two lemons and stir well. Let the milk set for 15 minutes. If the milk does not set, add more lemon juice.

DRAINING

1 hour 15 minutes - 2 hours 15 minutes	**Cheesecloth**	**Colander**

Pour the curds into a cheesecloth-lined colander. Tie the four corners of the cheesecloth into a knot and hang bag to drain for 1 to 2 hours or until the curds have stopped draining. Save the whey. It can be used in cooking, such as baking bread. Chilled and with mint leaves added, it makes a refreshing summer-time drink.

MIXING, SALTING, AND SPICING

Salt	**Herbs**

Take the cheese out of the cheesecloth. You may have to scrape some off the cloth. The cheese can be lightly salted and herbs may be added if desired. The yield should be about 6–8 ounces of lemon cheese for each quart of milk.

Neufchatel

This cheese originated in Normandy, France. It is a soft, milk cheese made from whole milk enriched with cream. The cheese may be eaten fresh. *Makes nearly 2 pounds.*

RIPENING AND RENNETING

12 to 18 hours	**72° F.** **1 pint heavy cream** **1 teaspoon diluted rennet**	**1 gallon whole milk** **4 ounces mesophilic cheese starter culture**

To 1 gallon of whole milk add 1 pint of heavy cream. Mix thoroughly. Indirectly warm the milk to 72° F. Add 4 ounces of mesophilic cheese starter culture. Add 1 teaspoon of a diluted rennet mixture.

(Dilution: 3 drops of liquid rennet in 1/3 cup cool, sterile water.) The exact amount of rennet is important. Too little rennet and the cheese will drain through the cheesecloth. Too much rennet will give the cheese a hard, rubbery texture. Let the milk set covered at 72° F. for 12 to 18 hours or until a thick curd has formed.

DRAINING

18 hours to 30 hours

Cheesecloth **Colander**

Pour the curds into a cheesecloth-lined colander and hang to drain for 6 to 12 hours or until the bag has stopped dripping.

2 to 2½ days

Cheesecloth **Colander**
Cheese pot

Place the curds into a cheesecloth-lined colander and place the colander in a pot. Place a plate in the colander, resting on the bag of curds. Place a weight on the plate (the weight of two bricks is sufficient). Put the cover on the pot and refrigerate for 13 hours.

MIXING, SALTING, AND SPICING

Salt **Spices**
Bowls **Saran Wrap or wax paper**

Take the cheese from the pot and place in a bowl. Knead and mold the cheese by hand into four cheeses. You can salt to taste and add a variety of condiments if desired such as chopped chives, chopped garlic cloves, chopped onions or scallions, or cut up pineapple, olives, or pickles. The possibilities are endless. Shape the cheeses and cover each with Saran Wrap or wax paper and store in the refrigerator. This cheese will keep a week.

Gervais

Gervais is a fresh French cheese which is made from a mixture of cream and milk. It is similar to Neufchatel but is richer and creamier. This is a fresh cheese which should be consumed within several days and for this reason it is rarely found in cheese stores. It is traditionally made with cow's milk, but can be made with goat's milk. *Makes 4–6 ounces.*

RIPENING AND RENNETING

1 day

65° F. **1 1/3 cups cream**
2 2/3 cups milk **2 ounces mesophilic cheese starter culture**
1 drop liquid rennet

Add 1 1/3 cups cream to 2 2/3 cups whole milk and mix thoroughly. Warm to 65° F. and add 2 ounces mesophilic cheese starter culture. Mix thoroughly. Add 1 drop of liquid rennet to 2 tablespoons water. Mix into the milk and cream and stir gently for 5 minutes. If the milk starts to coagulate, stop stirring.

Leave the milk to set for 24 hours by which time coagulation should be completed.

DRAINING

1 day 4 hours to 1 day 6 hours

Cheesecloth, fine weave
Colander
Ladle

Ladle the curds into a cheesecloth-lined colander. The cheesecloth must be of a fine weave. Tie the four corners of the cheesecloth together and hang the curds to drain for 4 to 6 hours or until the curds stop dripping. The bag of curds may have to be taken down every now and then to have the sides of the cloth scraped with a spoon to open the pores of the cloth for drainage.

PRESSING

Put the curds into fresh cheesecloth, place in a cheese mold, and press at 15 pounds pressure for 6 to 8 hours.

Remove the cheese from the press and place in a bowl; add a small amount of salt to taste. The cheese should be creamy and smooth. If it is not, press the cheese through a strainer. Herbs may be added now if desired. Place the cheese in small molds lined with wax paper. Traditionally these molds are 2 1/4 inches wide by 1 3/4 inches deep. One quart will make 4 cheeses.

Place in the refrigerator until ready for use. These cheeses will keep for up to 2 weeks.

Bondon

Bondon is another fresh French cheese. It is made from whole milk and is similar in taste and texture to Neufchatel. Traditionally, this cheese is made from cow's milk. However, with this recipe one can use goat's milk. *Makes 4–6 ounces.*

RIPENING AND RENNETING

24 hours

65° F.
1 quart whole milk
1 drop liquid rennet
2 ounces mesophilic cheese starter culture

To 1 quart of whole milk at 65° F. add 2 ounces of mesophilic cheese starter culture. Mix thoroughly. Add 1 drop liquid rennet to 2 tablespoons water and mix into the milk for several minutes. Cover and let the milk set at 65° F. for 24 hours.

DRAINING

1 day 6 hours to 1 day 8 hours	**Cheesecloth** **Ladle**	**Colander**

When the curd has coagulated, ladle it into a cheesecloth-lined colander. Tie the four corners of the cheesecloth into a knot and hang to drain for 6 to 8 hours or until the curds stop draining. The sides of the cheesecloth may need occasional scraping with a spoon to hasten draining.

PRESSING

Remove the curds from the cheesecloth and place into a cheesecloth-lined mold. Press the cheese for 4 to 8 hours at 15 pounds pressure.

Remove the cheese from the press and place in a bowl. Salt and spice to taste.

The cheese should be smooth in texture. If it is grainy, force it through a strainer. Place the cheese in small molds lined with wax paper. Traditionally these molds are 2 3/4 inches deep and 1 3/4 inches in diameter.

Place in the refrigerator until ready for use. These cheeses will keep for up to two weeks.

CREAM CHEESE

Cream cheese is a soft, rich cheese spread which is made from cream. There are a variety of recipes for this rich cheese. The two included here are for an uncooked curd cream cheese and a cooked curd cream cheese. *Both recipes make 6–8 ounces.*

Cream Cheese

Uncooked Curd Method

RIPENING AND COAGULATION

12 hours	**72° F.** **1/2 gallon cream**	**4 ounces mesophilic cheese starter culture**

To 1/2 gallon of fresh cream, add 4 ounces of mesophilic cheese starter culture. Let the cream stand for 12 hours at 72° F. By this time a solid curd should form.

DRAINING

24 hours **Colander** **Cheesecloth**

Pour the curd into a cheesecloth-lined colander, and tie the four corners of the cheesecloth into a knot. Hang the bag to drain for up to 12 hours or until the bag stops dripping. If the bag is changed every several hours, the dripping process will be speeded up.

MIXING, SALTING, AND SPICING

Salt **Herbs**

When the curd is drained, place it in a bowl, and add salt and herbs to taste. Place the cheese in small molds to give it shape, and cool it in the refrigerator. After gaining its shape, the cheese can be wrapped in Saran Wrap or wax paper and stored for up to a week in the refrigerator.

Cream Cheese

Cooked Curd Method

30 minutes **1/2 gallon cream** **145° F.**

Heat 1/2 gallon of cream to 145° F. for 30 minutes.

RIPENING AND RENNETING

12 hours 30 minutes **86° F.** **1 teaspoon diluted rennet** **4 ounces mesophilic cheese starter culture**

Cool the cream to 86° F. Add 4 ounces mesophilic cheese starter culture. Add 1 teaspoon of diluted rennet solution. (Dilution: 3 drops of liquid rennet to 1/3 cup of cool sterile water.) Let the cream set for 12 hours at room temperature (72° F.).

SCALDING AND DRAINING

18 hours 30 minutes **Water at 170° F.** **Cheesecloth** **125° F.** **Colander**

Pour the curded cream into a pot and add enough water at 170° F. to the curd to raise the temperature to 125° F. This may take from 1 quart to 1/2 gallon. Pour the curds into a colander lined with very fine cheesecloth. Tie the cheesecloth at the four corners and hang to drain.

MIXING, SALTING, AND SPICING

Salt **Herbs** **Saran Wrap or wax paper**

When the bag stops draining, place the cheese in a bowl, and add salt or herbs if desired. You can add a variety of condiments to cream cheese including chopped olives, onions, chives, and pineapple. Place the cheese in small molds and store in the refrigerator. Once the cheeses are firm they can be wrapped in Saran Wrap and stored in the refrigerator for up to a week.

Swiss Cream Cheese

This is an old recipe for a sweet cream cheese that is quite tasty. *Makes 4–6 ounces.*

RIPENING AND RENNETING

24 hours

65° F.
2 ounces mesophilic cheese starter
1 quart heavy cream
1 drop liquid rennet

To 1 quart of heavy cream at 65° F. add 2 ounces mesophilic starter culture. Mix in thoroughly. Add 1 drop liquid rennet to 2 tablespoons water. Stir into the cream and leave the cream to set at 65° F. for 24 hours.

DRAINING AND SALTING

36 hours

Cheesecloth
2 teaspoons salt
Colander

Line a colander with two layers of a fine cheesecloth. Pour half the coagulated cream into the cheesecloth-lined colander. Sprinkle on 1 teaspoon coarse cheese salt. Pour the remaining curded cream into the colander and sprinkle on the second teaspoon of salt. The salt will help in drainage. Tie the four corners of the cheesecloth and hang the bag to drain for 12 hours.

PRESSING

40–42 hours Place the curds in a cheesecloth-lined mold and press for 4 to 6 hours at 10 pounds pressure.

Remove the cheese from the mold. Place in a bowl. Add herbs if desired. Pack the cheese into small containers and refrigerate. The cheese should be eaten within several days.

French Cream Cheese

This is a soft, spreadable cream cheese that is quite sweet and requires little starter culture for ripening. *Makes 4–6 ounces.*

RENNETING

24 hours

70° F.
2 cups whole milk
2 cups cream
1 ounce mesophilic starter
1 drop liquid rennet

Mix 2 cups whole milk and 2 cups heavy cream. Add 1 ounce mesophilic starter and 1 drop liquid rennet. Stir. Allow to set at 70° F. for 24 hours.

DRAINING

30–36 hours

Cheesecloth
Colander

Ladle the curds into a cheesecloth-lined colander. Tie the four corners of the cheesecloth into a knot and hang to drain for 6 to 12 hours or until the curds stop dripping.

Place the curds in a bowl and mix by hand to a pastelike consistency. Chopped fresh herbs and salt to taste may be added. Additional heavy cream may be added if a moister cheese is desired.

The cheese can be pressed into individual serving containers and stored in the refrigerator until ready for use. French cream cheese will keep for several days under refrigeration.

COTTAGE CHEESE

Cottage cheese originated in eastern and central Europe. It is a soft, fresh, cooked curd cheese which is usually eaten within a week after being prepared. It was quite popular in colonial America and its name is derived from the fact that it was made in local cottages. This cheese is also known as farmer's cheese and pot cheese (because it was made at home in a pot). Several of the many variations in the preparation of this cheese are included in this chapter.

In days gone by this cheese was made from raw milk which was placed in a pot and set in a spot that would stay fairly warm. In winter this would be next to or on a cool corner of the cook stove. In several days, due to the action of bacteria present in the non-pasteurized milk, the lactic acid level would become so high that the milk protein would precipitate out in a soft, white curd. This soft curd could be treated in a variety of ways. It could be cut up, warmed to about 100° F. for several hours, and then drained to produce a tasty sour cottage cheese. Sometimes the curds were merely drained without cooking to produce a lactic acid-type cheese. Sometimes the curds would be pressed after cooking to produce what was called a farmer's cheese.

Small Curd Cottage Cheese

This cheese is coagulated by the action of the starter culture bacteria, instead of rennet. It has a pleasant, sour taste and will keep under refrigeration for 1 week. It is delicious when eaten alone, or can be used in recipes calling for cottage cheese. *Makes 1 1/2 pounds.*

RIPENING

1 day

72° F.
1 gallon skim milk

4 ounces mesophilic cheese starter culture

Warm 1 gallon of skim milk to 72° F. Stir in 4 ounces mesophilic cheese starter culture. Leave pot to set at 72° F. for 16 to 24 hours. The milk will have curdled by then, due to the lactic acid produced by the starter culture bacteria.

CUTTING THE CURD

1 day
15 minutes

72° F.

The curd will be softer than that of rennet-made curd. Cut the curd into 1/4-inch cubes and let it set for 15 minutes.

COOKING THE CURD

1 day
45 minutes

100° F.

Raise the temperature of the curd 1 degree a minute until the temperature reaches 100° F. Stir every several minutes to keep the curds from matting.

1 day
55 minutes

Hold the temperature at 100° F. for 10 minutes, stirring to keep the curds from matting.

1 day
1 hour
10 minutes

Raise the temperature to 112° F. over a 15-minute period (about 1 degree a minute).

1 day
1 hour
40 minutes

Hold the temperature at 112° F. for 30 minutes or until the curds are firm. The test for firmness is to squeeze a curd particle between the thumb and forefinger. If the curd is a custard consistency inside, it is not ready and should be cooked longer.

DRAINING

1 day
2 hours

When the curds are sufficiently cooked, let them settle to the bottom of the pot for 5 minutes. Pour off the whey. Line a colander with coarse cheesecloth and pour the curds into the colander. Let drain for several minutes. If a less sour cottage cheese is desired, the curds can be washed by dipping the bag of curds into a bowl of cool water.

Dip the curds several times and allow to drain for several minutes. Rinse the curds in a bowl of ice water to cool and place the bag of curds in a colander to drain for 5 minutes.

SALTING, CREAMING, AND SPICING

Place the curds in a bowl and break up any pieces that are matted together. Add several tablespoons of heavy cream to produce a creamier texture. Salt may be added to taste. Herbs may also be added or fresh fruit, such as cut pineapple, can be mixed in.

Large Curd Cottage Cheese

This cheese has slightly larger curds and is coagulated by the action of starter culture and a small amount of rennet. *Makes 1 1/2 pounds.*

RIPENING AND RENNETING

8 hours

72° F.
1 gallon skim milk
Rennet (tablet or liquid)
4 ounces mesophilic cheese starter culture

Warm 1 gallon of skim milk to 72° F. Add 4 ounces of mesophilic cheese starter culture and mix in thoroughly. Dissolve 1/4 rennet tablet in 2 tablespoons water (or 1/4 teaspoon liquid rennet in 2 tablespoons water). Add 1 tablespoon of the diluted rennet to the milk and mix in thoroughly. Let set for 4 to 8 hours at 72° F. or until the milk coagulates.

CUTTING THE CURD

72° F.

Cut the curd into 1/2-inch cubes. Allow to set undisturbed for 10 minutes.

8 hours 15 minutes

Warm the curds 2 degrees every 5 minutes until the temperature reaches 80° F. Stir so that the curds particles do not mat together.

8 hours 35 minutes

Warm the curds 3 degrees every 5 minutes until the temperature reaches 90° F.

8 hours 50 minutes Warm the curd 1 degree F. every minute until the temperature reaches 110° F., stirring to keep curds from matting.

9 hours 45 minutes Allow the curds to remain at 110° F. for 20 minutes or until curds are sufficiently cooked (no longer have a custard-like interior).

Continue the recipe as for small curd cottage cheese starting with draining on page 43.

Goat's Milk Cottage Cheese

This is a fine tasting cottage cheese which is made from whole goat's milk. It could be made from skim milk; however, you have to own a cream separator to get the cream out of goat's milk. A small amount of rennet must be added to the milk because the solid content of goat's milk is not sufficient to allow starter bacteria to adequately coagulate the curds. *Makes 1 1/2 pounds.*

RIPENING

18 hours

72° F.
1 gallon goat's milk

Rennet
4 ounces mesophilic cheese starter culture

Warm 1 gallon of goat's milk to 72° F. Stir in 4 ounces of mesophilic cheese starter culture. (A fresh goat cheese starter culture will give best results.) Dissolve 1 drop of liquid rennet into the milk and stir in thoroughly. Allow the milk to set at 72° F. for 12 to 18 hours until coagulated.

CUTTING THE CURDS

18 hours 15 minutes **72° F.**

The curds will be quite soft. Cut into 1/2-inch cubes. Allow the curds to settle for 15 minutes.

COOKING THE CURD

72°–90° F.

Heat the curd 3 degrees F. every 5 minutes for the next 30 minutes.

18 hours 45 minutes Heat the curds 1 degree F. a minute until the temperature is at 102° F.

19 hours 15 minutes Keep the curds at 102° F. for 30 minutes or until the curds are cooked firm and no longer have a custardy interior.

Continue the recipe as for small curd cottage cheese starting with draining on page 43.

HARD CHEESES

The purpose of making hard cheese is to transform the milk protein and butterfat (cream) of milk so that they will maintain their nutritional value, develop a pleasant taste, and remain preserved for months and even years longer than they could in fresh milk.

The steps involved are ripening the milk, renneting, cutting, cooking, and draining the curd, milling and salting, molding and pressing, drying and waxing, and aging.

During the making of a hard cheese, most of the water, milk sugar, and minerals are separated from the milk protein and butterfat.

The acid level in the milk must be raised during ripening. Cheese starter culture bacteria are added to the warm milk, and these bacteria consume milk sugar and produce lactic acid. This increase in acid level aids in the expulsion of whey from the curd, helps the rennet to coagulate the milk, helps preserve the final cheese, and aids in flavor development in the aging cheese.

The increase in acid level in the milk must proceed at the proper rate. You do not want too much or too little acid produced.

When a higher acid level has been established, rennet is added. This must be carefully measured out, diluted, and added to the ripened milk. Rennet will coagulate the milk protein (not the albuminous protein, which stays with the whey) into a solid mass of white curd. Trapped within this mass of curd is the butterfat and whey. The latter contains the water, albuminous protein, milk sugar, and minerals of the milk.

The next steps are to remove most of the whey from the curd without removing the butterfat. This is begun by cutting the curd into small pieces so that the whey will slowly drain from it.

The small curd pieces are warmed to 100° F. or higher in order to expel more whey. The curd is stirred during cooking to keep it from sticking together. It must be treated gently to avoid damaging the curd and losing butterfat. After cooking, the curd is drained of whey. This leaves a mass of curd which is broken into small pieces, salted, and placed in a cheese press for up to twenty-four hours to remove even more whey and give the cheese its final shape. The cheese is then air-dried, waxed, and aged under the proper conditions.

Let us go back over these steps in detail.

HARD CHEESES

Variety	*Amt. Milk*	*Type of Milk*	*Coagulation*	*Culture*	*Amt. of Cheese*	*Use of Cheese*
Cheddar	2 gal. whole	Cow's	Rennet ¼ tab., 1 tsp.	Mesophilic 2 oz.	2 lbs.	Table, cooking
Stirred Curd Cheddar	2 gal. whole	Cow's, Goat's	Rennet ¼ tab., ¾ tsp.	Mesophilic 2 oz.	2 lbs.	Table, cooking
Sage Cheddar	2 gal. whole	Cow's, Goat's	Rennet ¼ tab., ¾ tsp.	Mesophilic 2 oz.	2 lbs.	Table, cooking
Caraway Cheddar	2 gal. whole	Cow's, Goat's	Rennet ¼ tab., ¾ tsp.	Mesophilic 2 oz.	2 lbs.	Table
Jalapeno Cheddar	2 gal. whole	Cow's, Goat's	Rennet ¼ tab., ¾ tsp.	Mesophilic 2 oz.	2 lbs.	Table
Derby	2 gal. whole	Cow's, Goat's	Rennet ¼ tab., ¾ tsp.	Mesophilic 2 oz.	2 lbs.	Table, cooking
Leicester	2 gal. whole	Cow's	Rennet ¼ tab., 1 tsp.	Mesophilic 4 oz.	2 lbs.	Table, cooking
Gouda	2 gal. whole	Cow's	Rennet ¼ tab., 1 tsp.	Mesophilic 4 oz.	2 lbs.	Table, cooking
Caraway Gouda	2 gal. whole	Cow's	Rennet ¼ tab., 1 tsp.	Mesophilic 4 oz.	2 lbs.	Table
Hot Pepper Gouda	2 gal. whole	Cow's	Rennet ¼ tab., 1 tsp.	Mesophilic 4 oz.	2 lbs.	Table
Colby	2 gal. whole	Cow's	Rennet ¼ tab., 1 tsp.	Mesophilic 3 oz.	2 lbs.	Table, cooking
Swiss	2 gal. whole	Cow's	Rennet ¼ tab., ½ tsp.	Thermophilic 2 tbsp. Proprionic 1 tsp.	2 lbs.	Table, cooking

Variety	*Amt. Milk*	*Type of Milk*	*Coagulation*	*Culture*	*Amt. of Cheese*	*Use of Cheese*
Caraway Swiss	2 gal. whole	Cow's	Rennet ¼ tab., ½ tsp.	Thermophilic 2 tbsp. Proprionic 1 tsp.	2 lbs.	Table
Emmenthaler	2 gal. whole	Cow's	Rennet ¼ tab., ½ tsp.	Ementhaler ½ tsp. Proprionic 1 tsp.	2 lbs.	Table, cooking
Mozzarella	2 gal. whole	Cow's	Rennet ¼ tab., 1 tsp.	Thermophilic 4 oz.	2 lbs.	Cooking
Parmesan	2 gal. skim	Cow's, (Goat's)	Rennet ¼ tab., 1 tsp.	Thermophilic 4 oz.	2 lbs.	Grating
Parmesan, Piquant	2 gal.	1 gal. cow's skim; 1 gal. goat's whole	Rennet ¼ tab., 1 tsp.	Thermophilic 4 oz.	2 lbs.	Grating
Romano	2 gal. skim and cream	Cow's	Rennet ¼ tab., ¾ tsp.	Thermophilic 4 oz.	2 lbs.	Table or grating
Romano, Piquant	2 gal.	1 gal. cow's low-fat; 1 gal. goat's	Rennet ¼ tab., ¾ tsp.	Thermophilic 4 oz.	2 lbs.	Table or grating
Montasio	2 gal. whole	Cow's	Rennet ¼ tab., 1 tsp.	Mesophilic 1 oz. Thermophilic 2½ oz.	2 lbs.	Table or grating
Montasio, sharp	2 gal.	1 gal. cow's lowfat; 1 gal. goat's	Rennet ¼ tab., 1 tsp.	Mesophilic 1 oz. Thermophilic 2½ oz.	2 lbs.	Table or grating

Ripening

The first step in making a hard cheese is the ripening of the milk. The milk is warmed to a specified temperature, usually between 86° and 90° F. This is done indirectly by placing the pot of milk into a container or sink of warm water. When the milk has reached the required temperature, cheese starter culture is added. The starter contains active lactic acid-producing bacteria which will ripen (acidify) the milk, usually over a period of thirty minutes to one hour. The starter is usually mixed with a spoon to take out any lumps and then it is thoroughly mixed into the warm milk.

A good dairy thermometer is essential for keeping a close eye on the temperature of the ripening milk. If the temperature of the milk starts to get too high, take the pot out of the warm water. If the temperature falls too low, place the pot back into the warm water.

The type of starter used will depend on the temperature at which the curds are cooked. For moderate cooking temperatures such as for Cheddar, a mesophilic cheese starter culture will be used. For high cooking temperatures, such as for Parmesan, a thermophilic cheese starter culture will be used.

The increase in acidity at this stage is important, so you must keep a close eye on the temperature, time, and the amount of starter added. If too much acid is produced, the cheese will sour and could leak whey during aging. If too little lactic acid is produced by the

To make hard cheese, warm milk, usually to 86°–90°.

Next, stir cheese starter culture to remove lumps.

Add starter culture (and color, if desired) to milk.

Finally, add rennet, carefully diluted in cool water.

starter culture, the cheese may have little flavor and could have numerous gas holes due to the presence of contaminating yeast or coliform bacteria.

You should remember that in some ways cheese is alive. The starter culture bacteria must be healthy and active during cheesemaking. They must develop an increasing acidity in milk during the ripening stage or problems will ensue. Particular care must be exercised in the preparation and storage of your starter (see page 16.)

Coloring

If you desire to add cheese color to the ripened milk, do it before adding rennet. Cheese color can destroy the coagulating ability of rennet. The color should be diluted in twenty times its own volume of water in a separate container and thoroughly mixed into the milk.

Renneting

The next step is the addition of rennet. This, too, must be diluted in twenty times its own volume of cool water before being added to the ripened milk. Rennet tablets or liquid rennet may be used, and you must remember that it takes about ten minutes for a rennet tablet to dissolve. One-quarter cup of cool water is usually enough to dilute the rennet. If rennet is not diluted, it will be unevenly distributed in the milk, which could produce a faulty curd.

It is helpful to have a slotted spoon when adding the rennet. Pour it through the slotted spoon into the milk. Mix gently in an up-and-down motion for several minutes. If the milk being used is not homogenized, you should top-stir for several minutes to keep the cream from rising. Top-stirring is stirring the top 1/4 inch of the milk. The milk is then left to set for thirty to forty-five minutes or until the milk is firmly coagulated into a curd. A chemical reaction takes place in the milk and the enzyme rennin (or if using vegetable rennet, a microbial enzyme) makes the protein portion of the milk precipitate out of solution, giving a solid white custard-like material called curd. Trapped within this mass of white curd are the butterfat and whey, which will be separated in the following steps.

Top-stirring the milk for several minutes will keep cream from rising in early stages of coagulation.

Cutting the Curd

The curd is ready for cutting when it gives a clean break. This can be determined by placing a thermometer or clean finger into the curd at a forty-five-degree angle. If the curd separates very cleanly and clearly around the inserted thermometer, this is a clean break, and the curds are ready for cutting. If the curd does not show a clean break, wait another five minutes and test again.

Wait until curd gives a clean break before cutting it. If ready, curd will separate as shown, during testing.

The curd is cut according to the recipe and the

range is from 1/4-inch up to 1/2-inch cubes. A long-bladed knife which will reach to the bottom of the pot is needed. Let's say the curd is going to be cut into 1/2-inch cubes. Place your knife 1/2 inch from the left side of the pot and make a straight slice going all the way to the bottom of the pot. Make a second slice 1/2 inch to the right of your first slice. Continue making 1/2-inch slices all the way across the pot. Now you have a pot of curd cut into 1/2-inch slices.

Turn the pot 90 degrees, and repeat the cutting. When you are done you will have a checkerboard pattern of square curds.

Immerse a stainless steel ladle 1/2 inch below the surface of the curd. Move it across the pot at this level until the curds are cut. Lower the ladle another 1/2-inch and do the same thing. Continue to do this until you have cut all the curds.

Don't be alarmed if you do not cut each curd to the correct size. You simply want to have all of the curds as close as possible to the same size. Let the curds rest for five minutes to firm up and then stir them with the ladle, cutting any oversized pieces with the curd knife. Stir the curds very gently. Butterfat is trapped in them, and too harsh stirring will result in a loss of fat from the cheese. This will produce a cheese of poor consistency.

If the recipe calls for cutting the curd into 1/4-inch cubes, a stainless steel whisk can be used in place of a ladle to help cut the curds.

The curd is cut into uniformly sized small pieces to help remove most of the whey which is trapped in-

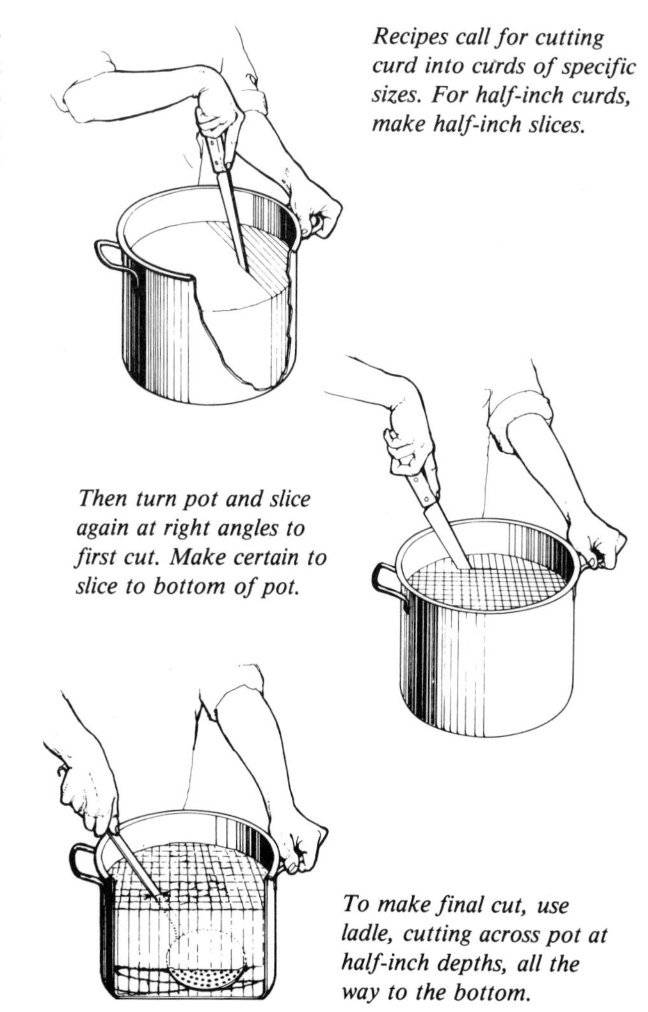

Recipes call for cutting curd into curds of specific sizes. For half-inch curds, make half-inch slices.

Then turn pot and slice again at right angles to first cut. Make certain to slice to bottom of pot.

To make final cut, use ladle, cutting across pot at half-inch depths, all the way to the bottom.

Cutting the curds requires these four steps. First, slice all of curds in one direction, using curd knife.

Now comes the most difficult step of process. Using ladle, cut curds at successive half-inch depths.

Next, slice curds at 90° angle to the first cut. Make certain curd knife tip reaches bottom of pot.

And finally, gently stir curds with the ladle. Cut any curds you find that are more than half-inch.

side it along with butterfat. Cutting the curd creates a huge increase in its surface area, and whey readily starts to drain away.

Cooking the Curds

After the curds have been cut into uniform-sized cubes you will note the cubes are floating in more liquid. After being cut, each cube slowly loses whey and shrinks in size. At this point the cubes are soft and jellylike, and must be handled very gently.

In making hard cheese you want to remove much of the moisture from the curds so that the cheese can be safely preserved. Raising the temperature of the curds helps in this process. The curds are usually cooked by placing a pot containing them in a container of warm water. The curds are gently stirred to keep from matting together. For many cheeses the temperature of the curd should rise no more than 2 degrees F. every five minutes. If the curds are warmed more quickly they may be damaged and will not drain properly. The temperature is controlled by moving the pot in and out of the warm water.

As the temperature slowly increases, the cubes will shrink in size since they are losing liquid whey. The amount of whey in the pot will noticeably increase. The cooking time the curds varies from thirty minutes to 2 1/2 hours. Cooking temperatures range from 100° to 132° F., depending on the particular cheese.

As curds are cooked they gradually lose whey content. Stirring during cooking keeps them from matting.

An automatic curd stirrer makes this task easier. Cheese recipes tell length of time for cooking curds.

Near the end of the cooking process, curds have shrunk to this size, and are much firmer in texture.

Once cooked, the curds are poured into a colander lined with cheesecloth and are hung up to drain.

As the curds are being cooked, the lactic acid-producing bacteria from the cheese starter culture are increasing the acid level in the curds and whey. You must cook the curds long enough so that the right amount of lactic acid is produced. You do not want to have too much acid produced or the cheese may be sour and bitter, with a moist soft texture. If too little acid is produced, you will produce a cheese with little flavor.

Draining the Curds

Once the curds have been cooked, drain them. Line a colander with coarse cheesecloth and pour the curds and whey into the colander. The whey may be saved for further cheesemaking. The curds are drained for five minutes up to an hour.

Milling and Salting

After draining, break the curds into small pieces ranging in size from a thumbnail to a walnut. This is called *milling* the curd. You will omit this step when making the Italian cheeses and Swiss cheese.

Sprinkle salt over the milled curds and gently but thoroughly mix it in. Salt helps bring out the flavor in the cheese and acts as a preservative. It also helps to expel more whey. Some cheeses such as Swiss or Gouda are not salted at this time; they are put into a brine bath later. It is best to use a coarse flake salt. It

Breaking up curds into small pieces can be child's play, as Ricki and Jenny Carroll demonstrate.

Sprinkle proper amount of salt over curds, then work it in. A coarse flake salt is recommended for all cheesemaking.

will mix into the curds more efficiently than a fine table salt.

Molding and Pressing the Cheese

Once the curds have been milled and salted, they are ready to be pressed. Place them in a form that will give the final cheese its shape. These forms are called cheese molds, and they may be made of wood, food grade plastic, or stainless steel. The cheese mold should not be made from soft lumber such as pine or oak which will impart a flavor to the cheese. Hard woods such as maple and birch are recommended. Tin cans should not be used as many of these are made with lead solder and this could be dangerous. The cheese mold aids in drainage of whey, so it should have holes along its sides.

The standard stainless steel cheese mold for a cheese made from two gallons of milk is an open-ended cylinder which measures 5 3/4 inches in diameter and 7 inches in height. Place the mold on a drip tray which will allow the whey to drain into a sink or small container. You must line the mold with a coarse cheesecloth before putting in the curds or you'll end up making a type of spaghetti, because the curds will be squeezed out the holes. After lining the mold with cheesecloth, quickly place the curds into the mold. You do not wish them to cool off.

Place a circular piece of cheesecloth cut to size on top of the curd, and a follower on top of that. The follower is usually made of wood and is needed to apply pressure to the curd. Sometimes a tight-fitting stainless steel follower will first be placed on the curd, followed by a looser wooden follower. Wood expands and contracts from the moisture and thus should not be made to fit snugly in the mold.

Once all the followers are in place, pull the cheesecloth liner snug to eliminate any bunching of the cloth, and lay it on top of the followers.

The cheese is now ready for pressing. A variety of presses are available. Lever presses are quite efficient, and if you are handy at woodworking, you can make up your own lever press. There are also a variety of home cheese presses available for purchase.

Line mold with cheesecloth, then place curds in mold, working quickly so curds don't become cool.

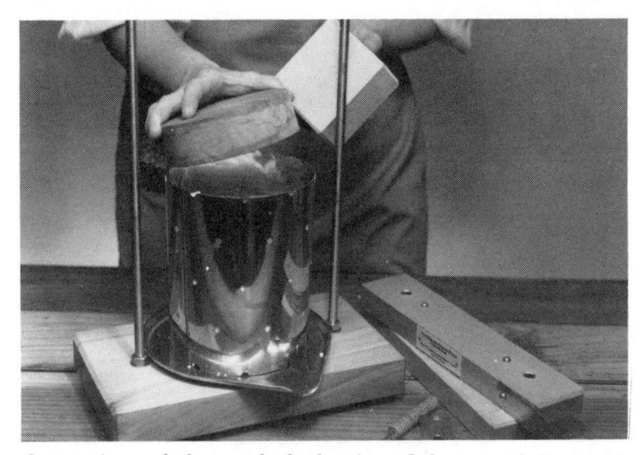

Cut a piece of cheese cloth the size of the top of the mold. Place it, then follower, on top of curds.

You should apply a light pressure, about ten pounds, to the cheese for the first fifteen minutes. The mold is then removed from the press and tipped upside down. Replace the followers and press the cheese at an increased pressure, usually twenty pounds, for thirty minutes. This flipping of the cheese will result in an even pressing. The cheese is usually flipped several times more and is finally left in the press at fifty pounds pressure for at least twelve hours.

For some cheese, you must change the cheesecloth each time the cheese is taken from the mold to increase the pressure. This is called redressing the cheese, and is necessary to keep the cloth from sticking to the curd.

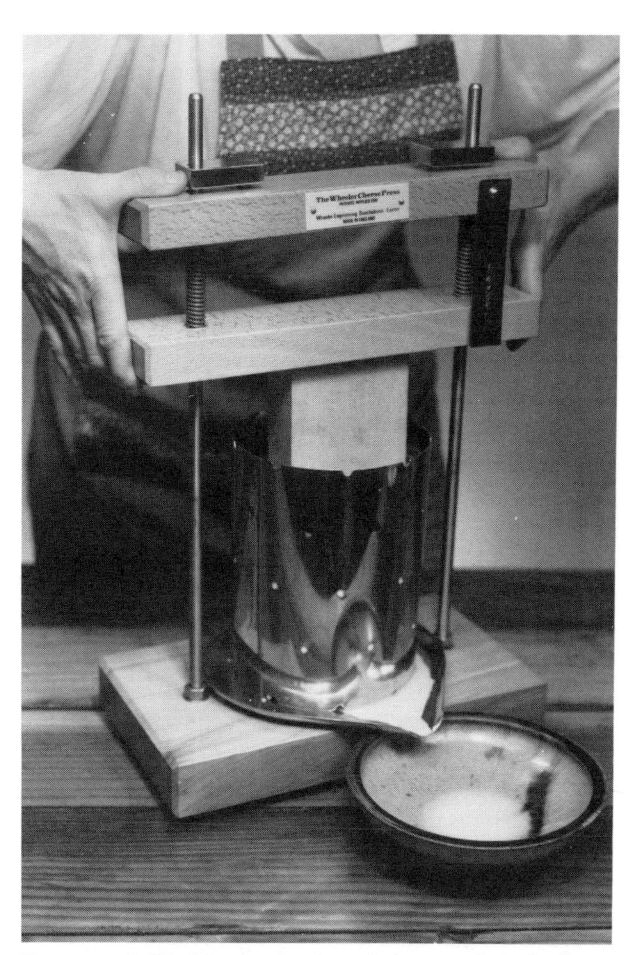

Pressure, outlined in most recipes, is increased gradually, and cheese is flipped several times.

Drying the Cheese

After pressing, remove the cheese from the press and gently peel off the cheesecloth. Place cheese on a cheese board or cheese mat and air-dry it at room temperature for several days. Turn the cheese several times a day to allow even drying of all surfaces. If any unwanted mold develops on the surface of the cheese, rub it off with a piece of cheesecloth dampened in vinegar.

Some cheeses such as Gouda or Swiss are placed in a brine solution immediately after coming out of the press. The brine is made by adding salt to a gallon of water, until the salt no longer dissolves. This takes about six cups of coarse salt. The container holding the brine must be noncorrosive. Glass or stainless steel is ideal. Placing the cheese in a brine adds salt to the cheese and produces a fairly tough rind on the maturing cheese. Brined cheeses are usually soaked from two to twenty-four hours.

Waxing the Cheese

Once the cheese is dry it may be waxed to keep the cheese from drying too much and to retard the growth of mold. Some cheeses such as Parmesan or Swiss have a thick natural rind and are not waxed.

Cool the cheese in the refrigerator for several hours prior to waxing, so that the wax will adhere better. A cheese wax is far preferable to any other type of covering since it gives a strong, flexible coating which will not crack as will the standard paraffin jelly wax.

Warm the wax in a double boiler on a stove that is vented with a stove hood fan. Wax vapors are flammable, so use caution in melting wax. Use a bristle brush to apply the wax. Wax one surface of the cheese and allow it to cool for several minutes before turning the cheese over to wax the other surface and sides. Two thin coats are preferable to one thick coat. Cheese wax dries very quickly. The second thin coat of wax can be applied as soon as the first coat dries. Write the name of the cheese and the date it was made on a piece of scrap paper, and wax this on the surface of the cheese, for accurate labeling.

Use bristle brush and cheese wax for best results.

Aging the Cheese

The cheese should be stored on a clean cheese board in a spot where the temperature will stay at 50° F. with a relative humidity of 85 per cent. Many home basements will satisfy this requirement. You may also want to purchase a second-hand refrigerator and use this as an aging room. Put a bowl of water in the refrigerator and keep the setting at 50° F. and you will have an ideal cheese storage chamber.

Turn the cheese over each day for the first several weeks and several times a week thereafter. This prevents moisture from accumulating on the bottom of the cheese.

The longer a hard cheese ages, the stronger its flavor becomes. A minimum of sixty days of aging is required for most cheeses. It is usually best to age a cheese for six months. Some of the hard grating cheeses can be aged for years and will develop a very sharp flavor.

Cheddar

Cheddar cheese is one of the most popular cheeses in the world. Its origins go as far back as the late 1500s, and its name is taken from the small village of Cheddar, in southern England. In colonial America it was one of the most common cheeses made by farm wives. The original recipes for Cheddar are somewhat involved and time-consuming; however, you will find the results to be well worth the effort. The cheeses included in this section all belong to the Cheddar family and all involve the same basic steps in their production. There are two recipes for Cheddar. The first includes the original process of *cheddaring* (cutting the drained curd into strips and allowing them to set at 100° F. for two hours). The second recipe is *stirred curd cheddar* and the cheddaring step is not needed, thus saving 2 1/2 hours of the cheesemaker's time. Other recipes included in the Cheddar family are Derby and Leicester, two English cheeses with a taste and texture very similar to Cheddar. *Makes 2 pounds.*

RIPENING

45 minutes	**86° F.** **2 gallons whole milk**	**2 ounces mesophilic cheese starter culture**

Heat 2 gallons of whole milk to 86° F. Stir in 2 ounces of mesophilic cheese starter culture. Cover and leave to ripen at 86° F. for 45 minutes.

RENNETING

1 hour 30 minutes	**86° F.**	**1/4 rennet tablet (or 1 teaspoon liquid rennet)**

Dissolve 1/4 rennet tablet or 1 teaspoon liquid rennet in 1/4 cup water. Gently stir the diluted rennet into the milk with an up-and-down movement for 1 minute. Top-stir the top 1/4 inch for several minutes. Cover the pot and allow to set undisturbed for 45 minutes.

CUTTING THE CURD

1 hour 35 minutes	**Curd knife**	**Ladle**

Cut the curd in 1/4-inch cubes. Allow the curds to set for 5 minutes.

COOKING THE CURD

2 hours 10 minutes Warm the curd to 100° F., increasing the temperature no more than 2 degrees every 5 minutes. Gently stir the curd to keep the particles from matting together.

2 hours 40 minutes Once the temperature has reached 100° F. (which should take about 35 minutes), maintain this temperature and stir the curds for 30 minutes.

3 hours Allow the curd to settle for 20 minutes.

DRAINING

3 hours 15 minutes **Colander** **Cheesecloth**

Pour the whey and curd into a colander. Place the colander of curd back into the pot and allow to set for 15 minutes.

CHEDDARING

5 hours 15 minutes Remove the colander from the pot and place the mass of curds on a cutting board. Cut the mass of curd into 3-inch slices. Place the slices in the pot in a sink where it rests in 100° F. water. Keep the curd slices at a temperature of 100° F.

First step in cheddaring is to cut curds into slices. These slices must be kept warm throughout process.

Turn them over every 15 minutes for 2 hours.

MILLING

5 hours 45 minutes Break the slices of curd into 1/2-inch cubes. The curd slices should be tough and have a texture similar to chicken meat. Leave the curd particles in the covered pot and place it back into a sink of 100° F. water. Stir the particles with your fingers every 10 minutes for 30 minutes. Do not squeeze the curd particles, merely stir them to keep them from matting together.

SALTING

Remove the pot from the sink. Add 2 tablespoons of coarse cheese salt. Gently stir it in.

PRESSING

Line a 2-pound cheese mold with cheesecloth. Place the curds into the mold. Press at 10 pounds pressure for 15 minutes.

2 days

Flip the cheese over and press at 40 pounds pressure for 12 hours. Flip the cheese over and press at 50 pounds pressure for 24 hours.

DRYING

4-7 days

Remove the cheese from the mold. Remove the cheesecloth. Air-dry cheese for 2 to 5 days or until it is dry to the touch.

WAXING AND AGING

Wax the cheese and put it to age at 55° F. for 3 to 12 months. The longer this cheese ages, the sharper the flavor it will develop.

Stirred Curd Cheddar

RIPENING

45 minutes

90° F.
2 gallons whole milk

2 ounces mesophilic cheese starter culture

Warm 2 gallons of whole milk to 90° F. Cow's or goat's milk may be used. Add 2 ounces of mesophilic cheese starter culture and stir thoroughly into the milk. Cover and leave the milk at 90° F. for 45 minutes.

COLORING

45 minutes

Cheese coloring

If cheese coloring is desired, add it now.

RENNETING

90° F.

1/4 cheese rennet tablet (or 3/4 teaspoon liquid rennet)

Dissolve 1/4 of a rennet tablet or 3/4 teaspoon of liquid rennet in 1/4 cup tap water. Stir the diluted rennet into the milk with a gentle up-and-down motion of the ladle for 1 minute. If using unhomogenized milk, top-stir the milk gently for several minutes to keep the cream from rising.

1 hour 30 minutes Let the milk set at 90° F. for 45 minutes or until the curd is firm and gives a clean break.

CUTTING THE CURD

1 hour 45 minutes

90° F. **Curd knife**
Ladle

Cut the curds into 1/4-inch cubes as uniform in size as possible. Let them set undisturbed for 15 minutes.

COOKING THE CURD

90°–100° F. **Ladle**

Stir the curds very gently. You do not want the curds to break apart from overstirring and you do not want the curds to mat together from lack of stirring.

2 hours 15 minutes Over the next 30 minutes warm the curds to a temperature of 100° F. Do not raise the temperature of the curds faster than 2 degrees every 5 minutes. Gently stir the curd.

2 hours 45 minutes

100° F.

Hold the temperature of the curds at 100° F. for an additional 30 minutes. Stir occasionally to keep the curds from matting together.

DRAINING

2 hours 50 minutes

100° F.

Drain the whey from the curds. Save the whey; it can be used in cooking or to make other cheeses. (See chapter on whey cheeses.) Drain by letting the curds settle for 5 minutes to the bottom of the pot and then pouring off most of the whey. Pour the curds into a large colander and further drain them for several minutes. Do not drain too long, or the curds will mat together.

STIRRING THE CURDS

100° F.

Pour the curds back into a pot and stir them briskly with your fingers, separating any curd particles that have matted together.

SALTING

3 hours Add 2 tablespoons of coarse cheese salt

to the curds. Mix in thoroughly. Do not squeeze the curds; simply mix the salt into them.

STIRRING

4 hours **100° F.**

Allow the curds to remain at 100° F. for 1 hour, stirring the curds every 5 minutes to avoid matting. The curds can be kept at 100° F. by resting the cheese pot in a sink or bowl of water at 100° F.

PRESSING

Line a 2-pound cheese mold with a piece of sterilized coarse cheesecloth. Place the curds into it. Add followers to the mold and press the cheese for 10 minutes at 15 pounds pressure.

1 day 6 hours Turn the cheese mold over and press at 30 pounds pressure for 10 minutes. Flip the mold over and press at 40 pounds pressure for 2 hours. Turn the mold over and press at 50 pounds pressure for 24 hours.

AIR-DRYING

3–7 days Remove the cheese from the press and gently peel off the cheesecloth. Place the cheese on a clean dry cheese board or cheese mat. Turn the cheese several times a day for several days until the surface of the cheese is dry to the touch. This takes from 2 to 5 days, depending on the humidity in your kitchen.

WAXING

Once the cheese is dry it can be waxed.

AGING

45° to 55° F. **85 percent relative humidity**

The cheese should be stored at 45° to 55° F. for 2 to 6 months. It should be cared for as has been described in the section on aging cheeses (page 60).

Sage Cheddar

You can use from 1 to 3 tablespoons of fresh chopped or dried sage. The amount depends on the degree of flavor you want in the final cheese. Place the sage in 1/2 cup water and boil for 15 minutes, adding water as needed so that it does not all boil away. Strain the flavored water into the milk to be used for cheesemaking. *Makes 2 pounds.*

Follow the directions for stirred curd Cheddar cheese. Add the sage during the salting.

Caraway Cheddar

This cheese is made by adding 1/2 to 2 tablespoons of caraway seeds to 1/2 cup boiling water. The seeds are boiled for 15 minutes, water is added so it all won't boil away, the water is strained into the milk, and the seeds are added during the salting process as described in the recipe for stirred curd Cheddar.

Jalapeno Cheddar

This cheese can be quite spicy. Add 1/2 to 4 tablespoons of chopped jalapeno peppers to 1/2 cup water and boil for 15 minutes, adding water as needed. Strain the water into the milk, follow the recipe for a stirred curd Cheddar, and add the peppers during the salting process.

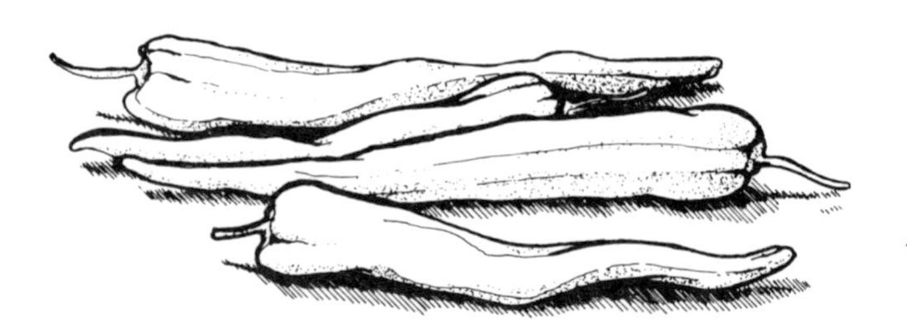

Derby Cheese

Derby originated in Derbyshire, England. It is very similar to Cheddar, but it has a higher moisture content and ages more quickly than Cheddar. *Makes 2 pounds.*

RIPENING

45 minutes	**84° F.** **2 gallons milk**	**2 ounces mesophilic cheese starter culture**

Heat 2 gallons of whole milk to 84° F. Add 2 ounces of mesophilic cheese starter culture and mix in thoroughly. Allow to set for 45 minutes.

RENNETING

1 hour 35 minutes	**84° F.**	**1/4 rennet tablet (or 3/4 teaspoon liquid rennet)**

Add 1/4 rennet tablet or 3/4 teaspoon liquid rennet to 1/4 cup tap water and gently stir this into the milk. Top-stir for 1 minute. Let set covered at 84° F. for 50 minutes.

CUTTING THE CURDS

	84° F.	**Curd knife**

Cut the curds into 1/2-inch cubes.

COOKING

2 hours 5 minutes

84°–94° F.

Gently stir the curds, and slowly raise the temperature to 94° F. over the next 30 minutes. Do not raise the temperature faster than 2 degrees every 5 minutes.

STIRRING

2 hours 15 minutes

94° F.

Stir the curds for 10 minutes.

DRAINING

2 hours 30 minutes

Pour the curds into a colander and allow to drain for 15 minutes.

3 hours 30 minutes

Place the mass of curds on a draining board and cut it into 2-inch slices. Lay the slices on the draining board and cover with a clean towel which should occasionally be placed in a bowl of water at 94° F., then wrung dry. You are trying to keep the temperature of the curd slices at 94° F. Turn the slices over every 15 minutes. Allow to drain for 1 hour.

MILLING AND SALTING

3 hours 30 minutes

2 tablespoons salt

Break the slices of curds into small pieces about the size of a quarter. The curd should be tough and tear when pulled apart. Mix 2 tablespoons of salt into the curds.

PRESSING

1 day 5 hours 40 minutes

Place the curds into a cheesecloth-lined mold. Press at 15 pounds pressure for 10 minutes. Flip the mold over and press at 30 pounds pressure for two hours. Flip the mold over and press at 50 pounds pressure for 24 hours.

AIR-DRYING

3–8 days

Take the cheese from the press and gently remove the cheesecloth. Place the cheese on a cheese board or cheese mat to dry for 2 to 5 days, turning the cheese several times a day.

WAXING

When the cheese is dry it may be waxed.

AGING

The cheese should be cured at 55° F. for 3 months, at which time it is ready to be eaten.

Leicester

This is a mild, hard cheese similar to Cheddar. It originated in Leicester County, England. The cheese ripens somewhat more quickly than Cheddar. It is traditionally made from cow's milk, and colored. *Makes 2 pounds.*

RIPENING

45 minutes | **85° F.** | **2 gallons whole milk** | **4 ounces mesophilic cheese starter culture**

Heat 2 gallons of whole milk to 85° F. Add 4 ounces of mesophilic cheese starter culture. Stir in thoroughly. Cover and allow to set for 45 minutes at 85° F.

COLORING

45 minutes | **85° F.** | **Cheese color**

Traditionally, cheese color is added at this point. One-eighth of a cheese color tablet dissolved in 1/4 cup of water and stirred in the milk will produce bright yellow cheese.

RENNETING

1 hour 30 minutes | **85° F.**

Add 1/4 rennet tablet or 1 teaspoon liquid rennet to 1/4 cup tap water. Gently stir in the dissolved rennet for several minutes. Allow the milk to set at 85° F. for 45 minutes.

CUTTING THE CURDS

1 hour 45 minutes | **85° F.** | **Curd knife**

Cut the curd into 1/4-inch cubes. Stir occasionally and gently for 15 minutes.

COOKING THE CURD

2 hours 40 minutes | **85°–95° F.**

Raise the temperature of the curd slowly to 95° F., allowing the temperature to rise 2 degrees every 5 minutes. Be sure to stir gently to avoid matting. Allow the curd to remain at 95° F. for 30 minutes, stirring to keep the curds from matting together.

DRAINING

4 hours | **Colander**

Drain the curds into a colander. Allow to drain for 20 minutes.

Pour the mass of curds from the colander onto a draining board. Cut into 2-inch slices to drain. Cover with a clean cloth which has been in 96° water and which has been wrung by hand so that it is not dripping wet. Do this to keep the curd slices warm. Turn them every 20 minutes for 1 hour.

MILLING AND SALTING

4 hours 10 minutes — **2 tablespoons salt**

Break the curd slices into pieces about the size of a nickel. Stir for several minutes. Add 2 tablespoons of coarse cheese salt and gently stir for several minutes.

PRESSING

1 day 6 hours

Place the curds into a cheesecloth-lined mold and press at 15 pounds for 30 minutes. Flip the mold over and press at 30 pounds for 2 hours. Flip the mold over and press at 50 pounds for 24 hours.

DRYING AND AGING

Take the cheese from the mold and gently remove the cheesecloth. Dust the cheese with coarse cheese salt and shake off any excess salt. Place the cheese on a cheese board and age where the temperature will remain at 55° F.

This cheese is ready to eat in 12 weeks.

WASHED CURD CHEESES

Both Colby and Gouda are washed curd cheeses. During cooking, whey is removed from the pot and is replaced by water. By washing the curd, the acid level is lowered. The milk sugar, or lactose, is washed from the curds to avoid souring the cheese. Such cheeses cure somewhat faster than other varieties and are ready for eating in twelve weeks. Colby and Gouda should not be aged for as long as a Cheddar.

Gouda Cheese

Gouda is a cheese of Dutch origin. It is washed curd cheese with a smooth texture and deliciously tangy taste. It looks particularly distinctive when covered with a traditional red cheese wax. *Makes 2 pounds.*

RIPENING

10 minutes — **90° F.**, **2 gallons whole milk**, **4 ounces mesophilic cheese starter culture**

Warm 2 gallons of whole milk to 90° F. Add 4 ounces of mesophilic cheese starter culture. Mix in thoroughly.

RENNETING

1 hour 15 minutes — **90° F.** — **1/4 rennet tablet (or 1 teaspoon liquid rennet)**

Dissolve 1/4 rennet tablet (or 1 teaspoon liquid rennet) in 1/4 cup cool water. Gently stir into milk for one minute. Top-stir 3 minutes. Cover and leave the milk to set for 1 hour.

CUTTING THE CURD

Stainless steel knife — **Ladle**

Using a stainless steel knife and ladle, cut the curds into 1/2-inch cubes. Treat the curd very gently.

COOKING THE CURD

100° F.

Raise the temperature of the curd to 100° F. during the next 30 minutes. Raise the temperature no faster than 2 degrees every 5 minutes. Stir the curd continuously and gently.

2 hours 15 minutes

Continue to stir the curd for the next 30 minutes, maintaining a temperature of 100° F. At 10-minute intervals remove 8 cups of whey from the pot and add 8 cups of water at 100° F. This should be done three times. Diluting the whey with water gives the cheese its typical smooth texture.

DRAINING

Pour off the whey and allow the curd to mat into a lump in the pot.

MOLDING AND PRESSING

5 hours 55 minutes — **Cheese mold**

Line a 2-pound cheese mold with cheesecloth. Add the curd, breaking it as little as possible. Add the followers and press at 20 pounds pressure for 20 minutes. Turn the mold over, replace the followers, and press at 30 pounds pressure for 20 minutes. Turn the mold over, replace the followers, and press at 40 pounds pressure for 3 hours. Remove from press.

SALTING

8 hours 55 minutes

Prepare a 20 percent brine by mixing 1 1/4 pounds of coarse salt in 1/2 gallon of cold water. This is about 3 cups of salt. Float the cheese in this brine for 3 hours. Make sure the container holding the brine is non-corrosive (glass or stainless steel).

DRYING

Remove the cheese from the brine. Pat dry with a paper towel or clean cheesecloth. Place on a cheese board or cheese mat and air-dry at 50° F. for 3 weeks. Turn the cheese daily. Remove any mold growth during this period by rubbing with cheesecloth dipped in vinegar.

WAXING

After drying, the cheese may be waxed. A red cheese wax adds a professional touch and will be appreciated by any who get to sample this homemade delicacy. Gouda may be aged for several months before being eaten. It should be stored at 50° F. and 85 percent relative humidity, and turned several times a week.

Caraway Gouda

Boil one tablespoon of caraway seeds in 1/2 cup water for 15 minutes, adding water as needed. Cool. Add the water to the milk before beginning the Gouda recipe. Add the seeds to the curds immediately after draining the curds. Gently mix in and then allow the curds to mat. The recipe for Gouda should be followed. *Makes 2 pounds.*

Hot Pepper Gouda

Hot peppers should be added to 1/2 cup boiling water. One teaspoon of dried jalapeno peppers will give a hot flavoring but your own taste buds will be the final judge as to amount. Boil the pepper for 15 minutes, adding water as needed. Add the water to the milk prior to cheesemaking. Add the peppers to the curds immediately after draining the curds. Gently mix in and then allow the curds to mat. The recipe for Gouda should be followed. *Makes 2 pounds.*

Colby

Colby cheese is a native American cheese named after a township in southern Wisconsin where it was first made. Colby is a type of Cheddar cheese in which the curds are washed during the cooking stage. It has a more open texture than Cheddar, contains more moisture, has a pleasant, mild flavor, and can be aged from 2 to 3 months. *Makes 2 pounds.*

RIPENING

1 hour	**86° F.** **2 gallons whole milk**	**3 ounces mesophilic cheese starter culture**

Warm 2 gallons of whole milk to 86° F. Add 3 ounces of mesophilic cheese starter culture. Mix in thoroughly. Let the milk ripen at 86° F. for 1 hour.

COLORING

1 hour	**86° F.**	**Cheese color**

If desired, cheese color may now be added. Dissolve 1/8 of a cheese color tablet in 1/4 cup of water, stir, and add to milk.

RENNETING

1 hour 30 minutes	**86° F.**	**1/4 rennet tablet (or 1 teaspoon liquid rennet)**

Dissolve 1/4 rennet tablet (or 1 teaspoon liquid rennet) in 1/4 cup cool water. Gently stir into the milk for 1 minute. Top-stir 3 minutes. Let the milk set for 30 minutes or until the curd shows a clean break.

CUTTING THE CURD

1 hour 40 minutes Cut the curd into 3/8-inch cubes. Stir gently. Let the curds rest for 5 minutes to firm up.

COOKING THE CURD

2 hours 50 minutes **86°–102° F.**

Raise the temperature of the curd 2 degrees every 5 minutes until the temperature reaches 102° F. Stir gently so that the curd particles do not mat together. Hold at 102° F. for 30 minutes, gently stirring the curds.

2 hours 55 minutes Let the curd set undisturbed for 5 minutes to settle on the bottom of the pot.

DRAINING AND WASHING

3 hours 10 minutes Drain off the whey to the level of the curd. Add tap water at 60° F. until the temperature of the curd and water reaches 80° F. Stir while adding the water. Hold the curd at 80° F. for 15 minutes. Stir to keep from matting. The moisture content of the cheese is controlled by the temperature of the water added. If a dryer cheese is desired, keep the curd-water mixture a few degrees higher than 80° F. If a moister cheese is desired have the temperature below 80° F.

DRAINING

3 hours 40 minutes Pour the curds into a cheesecloth-lined colander. Allow the curds to drain for 20 minutes.

MILLING AND SALTING

3 hours 50 minutes Break the curd into thumbnail-sized pieces. Add 2 tablespoons coarse flake salt and mix, thoroughly yet gently.

MOLDING AND PRESSING

17 hours 30 minutes Place the curd into a cheesecloth-lined mold. Press at 20 pounds pressure for 20 minutes, flip the cheese and press at 30 pounds for 20 minutes, flip the cheese again and press at 40 pounds for 1 hour, flip the cheese a final time and press at 50 pounds for 12 hours.

AIR-DRYING

Remove the cheese from the press. Remove the cheesecloth. Air-dry on a cheese board or mat for several days.

AGING

Wax and store at 50° F. for 2 to 3 months. Turn the cheese daily.

SWISS CHEESES

Swiss-type cheeses have been known since the time of the Romans. They were made in the Alpine mountains and are characterized by the holes found throughout them and by their fragrant, sweet, nutty flavor.

The flavor and holes (eyes) are not the result of mice; they are caused by bacteria, *Proprionibacterium*

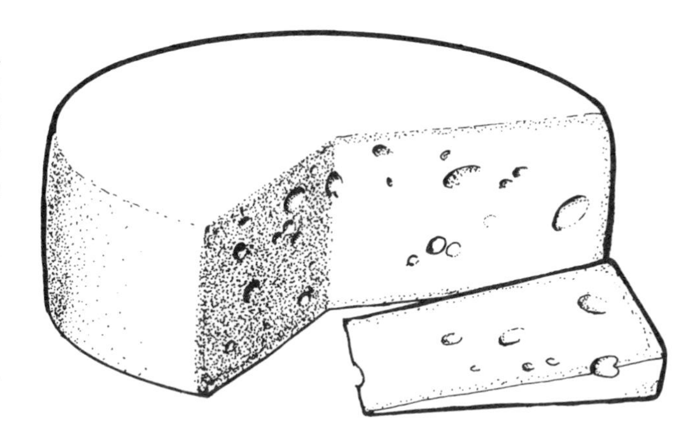

shermanii. These bacteria produce carbon dioxide gas which makes the eyes in the cheese. The bacterial starter for Swiss cheese consists of a 50/50 mix of *Lactobacillus bulgaricus* and *Streptococcus thermophilus.* These bacteria are referred to as thermophilic (heat-loving). During cooking, the temperature of the curds can go as high as 120° F. and thus heat-loving bacteria are needed to develop the lactic acid.

The cheese starter used for Emmenthaler cheese is also a thermophilic culture but it also contains the bacteria *Lactobacillus helveticus* which helps produce flavor.

The thick rind of Swiss cheese is a result of soaking the cheese in a brine prior to aging. The cheese is usually aged at least five months.

During the early part of aging, the cheese is stored in a warm room (70° F.) with a fairly high humidity (80 percent). The cheese is kept in such a room for two to three weeks in order for the Proprionic bacteria to grow and produce gas. *Makes 2 pounds.*

Swiss Cheese

RIPENING

10 minutes

90° F.
2 gallons whole milk
2 tablespoons thermophilic culture
1 teaspoon powdered Proprionic shermanii

Warm 2 gallons of whole milk to 90° F. Stir in 2 tablespoons of thermophilic culture. Take 1/4 cup of milk from the cheese pot and add 1 teaspoon powdered Proprionic bacteria to it. Mix thoroughly, making sure the powder is dissolved. Add to the milk and stir. Let set at 90° F. for 10 minutes to ripen.

RENNETING

40 minutes

90° F.
1/4 cup water
1/2 teaspoon liquid rennet (or 1/4 rennet tablet)

Dissolve 1/2 teaspoon liquid rennet (or 1/4 rennet tablet) in 1/4 cup water. Gently stir the dissolved rennet into the milk. If the milk is not homogenized, top-stir for several minutes. Leave the milk to coagulate for 30 minutes.

CUTTING THE CURD

50 minutes

Cut the curd into 1/4-inch cubes. This can be done by cutting the curd into 1/4-inch strips and then, with a pastry whisk, slowly stirring through the curds until each is about the size of a grain of rice. Do not stir vigorously or you will lose fat from the cheese. All curd pieces should be of equal size or too much moisture will be left in the cheese. Cutting will take about 10 minutes.

FOREWORKING

1 hour 30 minutes

90° F.

Keep the curd temperature at 90° F. Stir the curds for 40 minutes. This is called foreworking and helps to expel whey from the curds before they are heated.

COOKING

2 hours 35 minutes

90°–120° F.

Bring the temperature of the curds up to 120° F. in 30 minutes. This is done at a rate of 1 degree a minute. Maintain this temperature for 30 minutes, stirring often. The curds need to be cooked until they reach a stage called "the proper break." A test for the proper break is to wad a handful of curds together. Rub the wad gently between your palms. If the ball of curds readily breaks apart into individual curd particles, the curds are sufficiently cooked. Allow the curds to settle to the bottom of the pot for 5 minutes.

MOLDING AND PRESSING

2 hours 50 minutes

Cheese mold **Cheesecloth**

Pour off the whey. (Save for ricotta.) Have ready a freshly scalded 2-pound cheese mold with a coarse cheesecloth lining it. Quickly ladle the curds into the mold. You do not want the curds to cool. The mold should be in a sink or over a large pot so that the whey can easily drain. Fold a portion of the cheesecloth over the top of the curds, put a wooden follower in place, and add 8 to 10 pounds pressure. Let set 15 minutes.

3 hours 20 minutes

Cheesecloth

Turn the cheese over onto a fresh cheesecloth. Gently pull off the cheesecloth which covers the cheese. Lift the new cheesecloth up with the cheese in it and replace, upside-down, in the mold. Now that the cheese is in a fresh cloth and turned upside-down, press at 14 pounds pressure for 30 minutes.

5 hours 20 minutes

Turn the cheese onto a fresh cheesecloth and gently peel off the cloth around the cheese. Place the cheese with a new cheesecloth back into the mold for 2 hours, at the same pressure.

Change the cheesecloth and press at 15 pounds pressure for 12 hours.

SALTING

Day 2

Make up a saturated brine solution by placing 2 pounds of salt into 1 gallon of cold water. Stir the salt in thoroughly. Salt will be left on the bottom of the pot if the solution is saturated. The brine container must be non-corrosive, such as glass or stainless steel. Remove the cheese from the mold, gently peel off the cheesecloth, and place cheese in brine. Sprinkle salt on the surface of the floating cheese. Place the salt pot in the refrigerator. Leave in the brine for 12 hours.

CURING

Day 3

Remove the cheese from the brine. Place on a clean cheese board and store at 50°–55° F. with a high humidity (85 percent). Many basements will satisfy this requirement. A Styrofoam box with a bowl of water in it will also provide a humid environment. Turn the cheese daily for 1 week, wiping with a clean cheesecloth soaked in salt water. Do not wet the cheese.

10 days Place the cheese in a warm, humid room such as the kitchen (68°–74° F.). Turn daily and wipe with a cheesecloth soaked in salt water. Do not wet the surface of the cheese. The cheese can be left here for 2 to 3 weeks, until eye formation is noticeable. (The cheese will swell somewhat and become slightly rounded.)

4 weeks 3 days Place the cheese in a curing area where the temperature is 45° F. and the humidity is 80 percent. Allow to age 3 months or longer. Turn the cheese several times a week. Surface mold can be removed by rubbing with a cheesecloth soaked in salt water. A reddish coloration on the surface of the cheese is normal. This should not be removed.

Caraway Swiss

This is a tasty variation to the Swiss cheese recipe, and the caraway seeds impart their own distinct flavor to the cheese.

Place 2 tablespoons of caraway seeds in 1 cup of water. Boil in a small pot for 15 minutes. Strain out the caraway seeds. Add the water to the milk as the first step in the preparation of the cheese. Add the caraway seeds after draining the curds. Mix in thoroughly and quickly. You do not want the curds to cool before being placed in the cheese mold.

Follow the directions for the making of Swiss cheese. *Makes 2 pounds.*

Emmenthaler

This cheese originated in Switzerland. It has a sweet, nutty flavor. It calls for a culture known as Emmenthaler cheese starter culture. It is freeze-dried and contains *Lactobacillus bulgaricus* and *Streptococcus thermophilus,* plus *Lactobacillus helveticus,* which aids in flavor production. This culture can be added directly to the milk, 1/4 teaspoon of culture for each gallon of milk. *Makes 2 pounds.*

RIPENING

10 minutes

90° F.
2 gallons whole milk
1/2 teaspoon powdered Emmenthaler culture
1 teaspoon powdered Proprionic culture

Warm 2 gallons of whole milk to 90° F. Place 1/4 cup of the milk in a measuring cup and add 1/2 teaspoon of the powdered Emmenthaler culture and 1 teaspoon of the powdered Proprionic bacteria to the milk. Gently mix into the milk for a minute. Leave to ripen for 10 minutes.

RENNETING

40 minutes **90° F.** **1/4 rennet tablet (or 1/2 teaspoon liquid rennet)**

Dissolve 1/4 cheese rennet tablet (or 1/2 teaspoon liquid rennet) in 1/4 cup cool water. Gently mix into the milk and stir with up-and-down movements of the ladle for 1 minute. If using nonhomogenized milk, top-stir for several minutes. Let set for 30 minutes.

CUTTING THE CURD

50 minutes **Curd knife** **Stainless steel whisk**

Cut the curd into 1/4-inch cubes. This can be done using a curd knife to cut the curds into 1/4-inch cubes and a stainless steel whisk to break up the curd pieces until each is the size of a grain of rice. This should be done gently to avoid losing fat from the curd. Cutting should take 5 to 10 minutes.

FOREWORKING

1 hour 10 minutes Keep the curd particles at 90° F. for 20 minutes, gently stirring to keep the curd from matting together.

COOKING

1 hour 40 minutes **100° F.**

Slowly raise the temperature of the curds to 100° F. over the next 30 minutes. Stir to keep from matting. Raise the temperature 2 degrees F. every 5 minutes.

COOKING

2 hours 10 minutes **114° F.**

Raise the temperature of the curds to 114° F. during the next 30 minutes. Keep stirring. Raise the temperature 1 degree F. every 2 minutes.

2 hours 40 minutes Keep the temperature of the curds at 114° F. for 30 minutes or until the curds give a proper break.

DRAINING, MOLDING, PRESSING AND CURING

Continue the recipe as for Swiss Cheese on page 74.

ITALIAN CHEESES

The cheeses grouped in this section are the Italian cheeses which include mozzarella, Parmesan, Romano, and Montasio. All of these cheeses use a thermophilic (heat-loving) cheese starter culture in their preparation. Mozzarella, a soft, unripened cheese, is a pulled curd cheese in which the curds are heated to a very high temperature (170° F.) which gives the cheese its stretching quality.

Parmesan, Romano, and Montasio are hard cheeses which are called *grana* due to their granular texture. These cheeses are made with a low-fat milk (2.5 percent) and are usually aged for 10 months and consumed as a grating cheese. Most also may be aged for 3 months and can be used as a slicing cheese. *Makes 2 pounds.*

Mozzarella

Originally, mozzarella was made from buffalo's milk in Southern Italy. It is now made from cow's milk. This is an unusual cheese to make and involves a number of techniques unique to this type of cheese. The curds are heated to a very high temperature (170° F.) in order to produce the stretchiness that mozzarella is famous for. Thus you should have handy a pair of rubber gloves and wooden spoons to be used when working the mozzarella curd in the hot water. There are numerous recipes, including pizzas, which call for mozzarella in cooking, and thus it is one of the more useful and exciting cheeses to make at home. It is an excellent cheese for freezing and can be stored for several months in the freezer. *Makes 2 pounds.*

RIPENING AND RENNETING

45 minutes

90° F.
2 gallons whole milk
4 ounces Italian cheese starter culture (thermophilic)

1/4 rennet tablet (or 1 teaspoon liquid rennet) dissolved in 1/4 cup water

Warm 2 gallons of whole milk to 90° F. Add 4 ounces of Italian starter culture. Stir thoroughly. Dissolve 1/4 rennet tablet (or 1 teaspoon liquid rennet) in 1/4 cup water. Mix into the milk gently with an up-and-down stirring of the ladle for 1 minute. Leave the milk to set at 90° F. for 45 minutes or until the curd gives a clean break.

CUTTING THE CURD

1 hour

Cut curds into 1/2-inch cubes. Leave them in the pot for 15 minutes to firm up.

DRAINING

Drain the curds into a cheesecloth-lined colander. (Save the whey for ricotta.) Tie the four corners of the cheesecloth into a knot.

2 hours In a convenient spot for draining, hang up the bag of curds for an hour or until the curds stop dripping.

DEVELOPING ACIDITY

1 day 2 hours Place the bag in a colander and place the colander in a stainless steel pot. Put a cover on the pot and place it in the refrigerator for 24 hours. This will raise the acidity of the curds to a high level. Without this high level, the stretching quality of mozzarella will be missing.

WORKING THE CURD

After 24 hours, remove the bag from the colander and place the curds on a draining board or any convenient spot where they can be cut up and still drain some whey. To test to see whether the curds are ready for working, cut a 1/2-inch slice off the mass of curds and cut this slice into 1/2-inch cubes. Place the curds in a stainless steel bowl and cover them with 170° water.

First, cut mozzarella curds into half-inch cubes.

Place them in bowl and cover with 170° F. water.

Use two wooden spoons to work the curds by pressing them together into one another. Do this until the shape of the cubes no longer shows. If the curds become too hot, so they nearly melt, work them in a separate bowl. If too cool, raise the temperature of the water back to 170° F. You know that the curds are ready to work when the test ball lacks the shape of the cubes, the curd has a bright, shiny sheen, and it stretches easily. It will stretch under its own weight when you hold it. Place the test ball of curd in a bowl of ice water.

If the test curds did not behave properly, let the mass of curds rest for an hour and test again. If the curds are ready for working, cut the mass into 1/2-inch cubes and work about 1/2 pound of cheese at a time into a bright, shiny ball. Place each ball into the bowl of ice water to harden and to retain its shape. It should harden in 30 minutes.

If you wish a salted cheese, place the balls in a cold brine solution for 1 hour. The brine is made by dissolving 2 pounds of salt in 1 gallon of cold water.

Remove the cheese from the brine, dry it with a paper towel, wrap it in Saran Wrap, and refrigerate. Mozzarella will keep for up to 2 weeks in the refrigerator. It is also an excellent cheese to freeze.

Use two wooden spoons when working mozzarella curds.

Parmesan

Parmesan is a hard grating cheese which is usually made from low-fat (2.5 percent butterfat) cow's milk. The cheese originated in the region around Parma, Italy. Parmesan is usually aged for at least 10 months to develop a sharp flavor often described as piquant.

By mixing goat's milk with cow's milk, you can produce a stronger flavor in the cheese.

When using Parmesan, grate only the amount needed. The cheese will lose much of its flavor if grated, then stored. *Makes 2 pounds.*

RIPENING

30 minutes | **90° F.** | **2 gallons skim milk (2.5% fat)** | **4 ounces thermophilic cheese starter**

Warm 2 gallons of skim milk to 90° F. Add 4 ounces of thermophilic cheese starter culture. Allow to ripen for 30 minutes.

RENNETING

1 hour | **90° F.** | **1/4 rennet tablet (or 1 teaspoon liquid rennet)**

Dissolve 1/4 rennet tablet (or 1 teaspoon liquid rennet) in 1/4 cup cool water. Stir the diluted rennet into the milk and gently stir for 2 minutes. Allow to set for 30 minutes or until the curds give a clean break.

CUTTING THE CURD

1 hour 10 minutes | **90° F.** | **Curd knife** | **Stainless steel whisk**

Cut the curds into 1/4-inch cubes. Use a curd knife and stainless steel whisk.

COOKING THE CURD

1 hour 35 minutes | **90°–100° F.**

Raise the temperature of the curd 2 degrees every 5 minutes until the temperature reaches 100° F. Stir often.

2 hours 15 minutes | **100°–124° F.**

Raise the temperature of the curd 3 degrees every 5 minutes until the temperature of the curd reaches 124° F. Stir often.

2 hours 20 minutes | **124° F.**

The curds will now be about the size of a grain of rice. They will squeak if chewed. Allow the curds to settle 5 minutes.

DRAINING

2 hours 25 minutes

Pour off the whey without losing any of the curd particles.

MOLDING AND PRESSING

Line a 2-pound cheese mold with a

coarse cheesecloth. Pack the curds into the mold and press lightly, 5 pounds, for 15 minutes.

2 hours 40 minutes Remove the cheese from the mold and gently peel off the cheesecloth. Place the cheese on a fresh cheesecloth and replace into the mold. Reverse the cheese so that the top is now on the bottom. Press at 10 pounds for 30 minutes.

3 hours 10 minutes Remove the cheese from the mold, place in a fresh cloth, and put back into the press at 15 pounds pressure for 2 hours.

5 hours 10 minutes Remove the cheese from the mold, place in a fresh cloth, and put back into the press at 20 pounds pressure for 12 hours.

Day 2 Remove the cheese from the press. Remove the cheesecloth. Place the cheese in a saturated salt solution (2 pounds of salt in 1 gallon of water) for 24 hours.

Day 3 Remove the cheese from the brine. Place on a clean cheese board and cure it at 55° F. and 85 percent relative humidity for 10 months or longer. Turn the cheese daily for the first several weeks. Remove any mold growth. After 2 months of curing, rub the surface of the cheese with olive oil.

Parmesan

Piquant, Sharp

For a stronger flavored Parmesan, use 1 gallon cow's milk (2.5% fat) and 1 gallon goat's milk. Follow the recipe for Parmesan. *Makes 2 pounds.*

Romano

Romano is hard Italian grating cheese. If aged for several months it may be used as a table cheese. Longer aging (over 6 months) will produce a grating cheese. *Makes 2 pounds.*

RIPENING

10 minutes **88° F.** **2 gallons skim milk (2.5 percent fat)** **6 ounces cream** **4 ounces thermophilic cheese starter culture**

Warm 2 gallons of skim milk to 88° F. Mix in 6 ounces heavy cream. Mix in 4 ounces of thermophilic cheese starter culture. Allow the milk to ripen at 88° F. for 10 minutes.

RENNETING

40 minutes **88° F.** **1/4 rennet tablet (or 3/4 teaspoon liquid rennet)**

Dissolve 1/4 rennet tablet (or 3/4 teaspoon liquid rennet) in 1/4 cup cool water. Gently stir into the milk for several minutes. Allow to set until curd shows a clean break.

CUTTING THE CURDS

50 minutes Using a curd knife and stainless steel whisk, cut the curds into 1/4-inch cubes.

COOKING THE CURD

Raise the temperature of the curd to 116° F. during the next 45 minutes. Heat slowly at first (2 degrees every 5 minutes).

1 hour 35 minutes Gradually increase the heating time to 1 degree per minute.

2 hours 5 minutes **116° F.**

Keep the curds at 116° F. for 30 minutes or until the curds become firm enough so that when they are squeezed they retain their shape.

DRAINING

2 hours 10 minutes Drain off the whey.

MOLDING AND PRESSING

Day 2 Line a 2-pound cheese mold with cheesecloth. Place the curds into the mold. Press at 5 pounds pressure for 15 minutes. Remove the cheese from the press and place on a fresh cheesecloth. Turn cheese over and place back into the mold. This is called redressing. Press at 10 pounds for 30 minutes. Redress the cheese in a clean cheesecloth and press at 20 pounds for 2 hours. Redress the cheese in a clean cheesecloth and press at 40 pounds for 12 hours.

BRINING

Remove the cheese from the press. Gently remove the cheesecloth. Place the cheese in a saturated brine solution (2 pounds salt in 1 gallon of water) for 12 hours. Keep the brine and cheese in the refrigerator.

Day 3 Place the cheese on a clean cheese board and age at 55° F. with a relative humidity of 85 percent for 5 to 12 months.

Turn the cheese daily and remove any mold with a cloth dipped in vinegar. The cheese may be lightly rubbed with olive oil after two months to develop a less dry cheese.

Romano

Piquant, Sharp

For a stronger flavored Romano use 1 gallon of low-fat cow's milk (2.5 percent fat) and 1 gallon of goat's milk. Follow the recipe for Romano. *Makes 2 pounds.*

Montasio

Montasio is a hard Italian cheese which may be used as a table cheese when aged for 3 months and as a grating cheese if aged longer. This was originally a monastery cheese. It can be aged for 3 months and used as a table cheese or it can be aged for close to a year and used as a grating cheese for table or cooking. *Makes 2 pounds.*

RIPENING

1 hour — **88° F.** — **2 gallons whole milk** — **1 ounce mesophilic cheese starter culture** — **2 1/2 ounces thermophilic cheese starter culture**

Warm 2 gallons of whole milk to 88° F. Add 1 ounce of mesophilic starter culture and 2 1/2 ounces of thermophilic cheese starter. Mix in thoroughly. Allow to ripen at 88° F. for 60 minutes.

RENNETING

1 hour 30 minutes — **88° F.** — **1/4 rennet tablet (or 1 teaspoon liquid rennet)**

Dissolve 1/4 rennet tablet (or 1 teaspoon liquid rennet) in 1/4 cup cool water. Stir in gently but thoroughly for several minutes. Allow the curds to set 30 minutes or until the curds give a clean break.

CUTTING THE CURDS

1 hour 45 minutes — **Curd knife** — **Whisk**

Cut the curds into 1/4-inch cubes using a curd knife and stainless steel whisk.

COOKING THE CURDS

3 hours 20 minutes — **88°–102° F.**

Raise the temperature of the curd 2 degrees every 5 minutes until a temperature of 102° F. is reached. Stir often to keep the curds from matting together. Maintain this temperature for the next 60 minutes, and continue stirring.

DRAINING THE CURDS

3 hours 30 minutes **102°–110° F.**

Drain off the whey to the level of the curds. Add hot water until the curd-whey mixture is at 110° F. Maintain that temperature for 10 minutes. Stir the curds to keep from matting. Drain off the whey.

MOLDING AND PRESSING

Day 2 Line a 2-pound cheese mold with cheesecloth. Quickly place the curds into the mold. Press at 5 pounds pressure for 15 minutes. Turn the cheese out onto a fresh cheesecloth. Remove the old cloth, turn the cheese over, and place it back into the mold with its new cloth. Press at 5 pounds pressure for 1/2 hour. Turn the cheese out onto fresh cloth and replace into the mold. Press at 10 pounds pressure for 12 hours.

BRINING

Place the cheese into a saturated brine solution (2 pounds salt in 1 gallon of water) for 6 hours.

AGING

Place the cheese on a cheese board and keep at 55°–60° F. for 2 months. The relative humidity should be close to 85 percent.

Montasio

Sharp

For a stronger flavored Montasio, use 1 gallon low-fat cow's milk mixed with 1 gallon goat's milk. *Makes 2 pounds.*

WHEY CHEESES

When you make cheese there is usually a large amount of whey left over. The whey is the clear, greenish liquid which is separated by cooking and draining from the curds. It contains the milk sugar (up to 5 percent), albuminous protein, and minerals of the milk. Whey can be used in cooking and is particularly delicious if used in baking breads. It can also be used as a soup stock. It makes a refreshing summertime drink if served with ice and crushed mint leaves. It is also an excellent feed for pets, and livestock such as chickens or pigs.

A number of cheeses can be made from whey that is not over an hour old. The whey may be heated over a direct heat source.

Mysost

This cheese originated in the Scandinavian countries and is made from the whey of cow's milk. It has a unique sweet-sour flavor and is often served on hot toast for breakfast. The color ranges from light brown to dark brown depending on the amount of caramelization of the sugar and whether cream has been added. *Makes 1 1/2 pounds.*

BOILING DOWN

6–12 hours **The whey from the making of a cheese using 2 gallons of milk.** **One to 2 cups heavy cream**

Place fresh whey from the making of a 2-gallon batch of cheese into a pot. Add 1 to 2 cups of heavy cream. The amount added will determine the final texture of the cheese. If no cream is added, the cheese will be dark brown and have a slightly grainy texture. With the addition of cream, the cheese will be a light tan and the final texture will be somewhat smooth. Bring the pot of whey to a boil. Use of a wood cook stove is very economical in making this cheese since many hours of boiling are involved. Watch the pot carefully. As soon as the

WHEY CHEESES

Variety	*Amt. of Whey and milk*	*Type*	*Coagulation*	*Amt. of Cheese*	*Use of Cheese*
Mysost	Whey from cheese recipe using 2 gal. of milk.	Cow's	Lactic acid + Heat	1½ lb. +	Table or spread
Gjetost	Whey from cheese recipe using 2 gal. of milk.	Goat's	Lactic acid + Heat	1½ lb. +	Table or spread
Ricotta	2 gal. whey 1 qt. whole milk.	Cow's, Goat's	¼ cup vinegar	1-2 cups	Cooking
Ziegerkase	2 gal. whey 1 qt. whole milk.	Cow's	¼ cup vinegar	6-8 oz.	Table

whey begins to boil, a foam will appear on the surface. Remove this with a slotted spoon. The foam may be saved in a bowl, kept refrigerated, and added later. If the foam is not removed, the whey will boil over.

The whey needs to boil slowly uncovered over a low heat. When it is down to 75 percent of its original volume (this can take 6 to 12 hours), stir it so that it does not stick to the bottom of the pot. The reserved foam can now be added.

BLENDING

When the whey starts to thicken, place it in a blender and blend it at a high speed for a short time until its consistency is smooth. The whey is quite hot. Use caution when placing it in the blender.

BOILING

Pour the blended mixture back in the pot and continue to boil over a low heat, stirring continuously. The mixture will thicken.

MOLDING

When it approaches a fudge-like consistency, place the pot in a sink of cold water and stir the whey continuously until it is cool enough to be poured into molds. If the whey is not stirred, the cheese may be grainy. Once cool, it can be removed from the mold and covered with Saran Wrap or wax paper and stored in the refrigerator. If a more spreadable consistency is desired, shorten the boiling time somewhat. If a cheese that can be sliced is desired, heat the whey to a thicker consistency before molding.

For a variation, add crushed walnuts to the thickened whey just before it is cooled.

Gjetost

This cheese is made with the whey from goat's milk. Goat's cream may be added to the whey for a smoother cheese. The directions for making this cheese are exactly the same as for Mysost. The cheese has a tan color and a unique sweet-sour flavor. *Makes 1 1/2 pounds.*

Ricotta

Ricotta cheese originated in Italy. This is a slightly grainy, soft, fresh cheese which is used extensively in Italian cooking. Fresh whey must be used or the recipe will fail. *Makes 1–2 cups.*

HEATING

200° F.
1 quart whole milk (optional)
2 gallons fresh whey (no more than 1 hour old)

You may add 1 quart of whole milk to the fresh whey for an increased yield of cheese. Place 2 gallons of fresh whey in a pot on the stove. Bring the temperature of the whey to 200° F.

RIPENING

200° F.
1/4 cup cider vinegar

While stirring, turn off the heat and add 1/4 cup vinegar. You will notice tiny particles of white which is the precipitated albuminous protein. Line a colander with cheesecloth of very fine weave (butter muslin quality). Carefully pour the pot of whey into the colander.

Allow it to drain. When the cheesecloth is cool enough to handle, tie the four corners of the cloth into a knot and hang to drain for several hours. When the cheese stops draining, place it in a bowl and add salt and herbs to taste. Add a small amount of cream to give a richer, moister cheese. Several ounces of mesophilic starter culture may be added to enhance the flavor.

This recipe has a low yield. From 2 gallons of whey you can expect 1 to 2 cups of cheese.

The cheese should be refrigerated until used. It will keep up to a week.

Ziegerkase

This whey cheese originated in Germany. It may be eaten fresh or aged for several weeks. *Makes 6–8 ounces.*

HEATING AND RIPENING

30 minutes

200° F.
1 quart whole milk (optional)
1/4 cup coarse cheese salt
2 gallons fresh whey
1/4 cup vinegar
1 quart wine

You may add 1 quart of whole milk to the fresh whey for an increased yield of cheese.

Heat 2 gallons of fresh whey to 200° F. Slowly mix in 1/4 cup vinegar. For a variation, a herbed vinegar may be used. Turn off the heat. Allow the whey to set 10 minutes. You should see white flakes of protein floating in the whey. Line a colander with a fine weave cheesecloth (butter muslin quality) and pour the whey into it.

DRAINING AND PRESSING

1 day

Allow to drain. When the cheesecloth is cool enough to handle, tie the four corners into a knot and hang to drain for several hours or until the curds stop dripping whey. Allow the curds to cool. Line a 1-pound cheese mold with cheesecloth. Place the curds into it and press at 20 pounds pressure for 24 hours.

FLAVORING

5 days

Remove the cheese from the press and gently remove the cheesecloth. Mix 1 quart of wine, 1 quart of water, and 1/4 cup coarse cheese salt in a bowl. Place the cheese in the bowl and cover with Saran Wrap. Place the bowl in the refrigerator for 4 days, turning the cheese twice a day. Herbs may be added to the bowl to give an added flavor to the cheese.

AGING

Remove the cheese from the water-wine brine. Dry on a paper towel and cover with Saran Wrap. The cheese may be eaten fresh or aged for several weeks in the refrigerator before eating.

GOAT'S MILK CHEESES

There are subtle differences between cow's milk and goat's milk that the cheesemaker should recognize. A goat's milk cheese will be bright white, unless you add color, because the milk does not contain carotene. If serving a white Cheddar troubles you, you should add cheese color to the milk immediately after the ripening period. This will not change the flavor or texture of the cheese.

Goat's milk also contains a number of fatty acids that can develop a sharp flavor in cheese. For this reason you should use only fresh goat's milk in cheesemaking. The milk should taste sweet and have no noticeable flavor. When making Cheddar-type hard cheeses, the cheese starter should be a mesophilic culture which contains a minimum of flavor-producing bacteria. Purchase a mesophilic goat cheese starter culture from any of the cheesemaking supply companies listed in the appendix. For the Swiss and Italian cheeses, you can use the same thermophilic culture that is used with cow's milk.

When comparing the curds of the two milks, you will notice that goat's milk curd is somewhat softer and therefore must be treated quite gently. After cutting the curd, you may have to allow the curds to settle for 10 minutes to firm up enough to start the cooking process.

Almost all of the recipes in this book can be made using goat's milk. The recipes in this chapter, however, are specifically for goat's milk and include a soft goat's milk cheese, Saint Maure (a soft mold-ripened goat's milk cheese), a whole milk goat ricotta, Feta, and a goat's milk Cheddar.

Soft Goat Cheese

This is a delicious, soft goat's milk cheese. The milk is ripened for a lengthy period with goat cheese starter culture. A very small amount of rennet is also added to the milk. After 18 hours the milk coagulates. It is placed in small goat cheese molds to drain and in 2 days small and delicious 1 1/2- to 2-ounce cheeses are ready for eating. These are firm yet spreadable cheeses which will keep under refrigeration up to 2 weeks. *Makes almost 1 pound.*

GOAT'S MILK CHEESE

Variety	*Amt. Milk*	*Coagulation*	*Culture*	*Amt. of Cheese*	*Use of Cheese*
Soft Goat Cheese	1/2 gal.	Rennet 1/5 drop	Mesophilic goat 1 oz.	3/4 lb. +	Table
Herbed Cheese	1/2 gal.	Rennet 1/5 drop	Mesophilic goat 1 oz.	3/4 lb. +	Table
Saint Maure	1/2 gal.	Rennet 1/5 drop	Mesophilic goat 1 oz.	3/4 lb. +	Table
Feta	1 gal.	Rennet 1/4 tab., 1/2 tsp.	Mesophilic goat 2 oz.	1 lb.	Salads
Cheddar	2 gal.	Rennet 1/4 tab., 1 tsp.	Mesophilic goat 2 oz.	2 lb.	Table, cooking
Ricotta	2 gal. whey, + 1 qt. milk	1/4 cup vinegar	—	4–8 oz.	Cooking, desserts

RIPENING AND RENNETING

18 hours

72° F.
1/2 gallon whole goat's milk

1 ounce mesophilic goat cheese starter culture
liquid rennet

Warm 1/2 gallon whole goat's milk to 72° F. Stir in 1 ounce of mesophilic goat cheese starter culture. Place 5 tablespoons of cool water in a measuring cup. Add 1 drop liquid rennet and stir. Add 1 tablespoon of this diluted rennet to the milk. Stir thoroughly.

Cover and allow the milk to set at 72° F. for 18 hours, until it coagulates.

MOLDING AND DRAINING

72° F.

Scoop the curd into individual goat cheese molds. These molds are of food-

After milk has coagulated, scoop it into these individual plastic goat cheese molds, for draining.

Cheese must drain for two days, and will sink to one inch in height. Place it on rack over pan.

grade plastic and measure 3 1/4 inches in height. When the molds are full they should be placed to drain in a convenient spot.

Day 3 After 2 days of draining the cheese will have sunk down to about 1 inch in height and will maintain a firm shape. The cheeses can now be eaten fresh or can be wrapped in cellophane (better) or Saran Wrap and stored for up to two weeks in the refrigerator. If desired the cheese may be lightly salted on its surface, immediately after being taken from the mold.

Herbed Soft Goat Cheese

Follow the directions for making a soft goat cheese. When you scoop the curd into the cheese molds, sprinkle in layers of herbs. Chopped garlic, onion, and paprika make a tasty combination. Dill seeds, caraway seeds, or fresh ground black pepper can be added separately or in various combinations to spice up this cheese. *Makes almost 1 pound.*

Saint Maure

This is a soft mold-ripened goat cheese. The white mold (*Penicillium candidum*) is grown on the surface of the cheese and helps produce its pungent flavor. *Makes almost 1 pound.*

Follow the directions for producing the soft goat milk cheese. When the cheeses are removed from the molds, lightly salt all surfaces of the cheese. Lightly spray with a solution of white mold powder *(Penicillium candidum).* Place the cheese to age at 50° F. with a relative humidity of 95 percent for 14 days. By this time the cheese will have developed a thick coat of white mold. The cheese can now be wrapped in Saran Wrap and stored under refrigeration for up to two weeks. For a further discussion of the application of white mold powder, see the next chapter, on soft mold-ripened cheeses.

Goat cheeses are sprayed with white mold powder.

Feta

Feta is a heavily salted cheese which has its origins in Greece and was made from sheep's milk or goat's milk. It is often broken up into small pieces and used to garnish fresh salads. *Makes 1 pound.*

RIPENING

1 hour	**86° F.** **1 gallon whole milk**	**2 ounces mesophilic goat cheese starter culture**

Warm 1 gallon of whole goat's milk. Add 2 ounces of mesophilic goat cheese starter culture and mix in thoroughly. Allow to ripen for 1 hour.

RENNETING

2 hours	**86° F.**	**1/4 rennet tablet (or 1/2 teaspoon liquid rennet)**

Dissolve 1/4 rennet tablet (or 1/2 teaspoon liquid rennet) in 1/4 cup cool water. Stir gently into the milk for several minutes. Cover and allow to set for 1 hour.

CUTTING THE CURD

2 hours 10 minutes

86° F.

Cut the curd into 1/2-inch cubes. Allow to set undisturbed for 10 minutes.

86° F.

Gently stir the curd for 20 minutes.

DRAINING

Line a colander with cheesecloth. Pour the curds into the colander. Tie the four corners of the cheesecloth into a knot and hang to drain for 4 hours.

6 hours 30 minutes

Take down the bag and slice the curd into 1-inch slices, then cut the slices into 1-inch cubes. Sprinkle the cubes with 4 tablespoons of coarse flake salt. Place the cheese in a covered bowl and allow to age for 4 to 5 days in the refrigerator.

If a stronger-flavored cheese is desired, the cheese may be stored in a brine solution under refrigeration for 30 days. The brine is made by adding 2 1/4 ounces (1/3 cup) of coarse salt to 1/2 gallon of water.

Goat's Milk Cheddar

This is a sharp, peppery goat's milk cheese which is a stirred curd variety of Cheddar. It can be consumed after aging for 4 weeks but improves with flavor if aged up to 12 weeks. *Makes 2 pounds.*

RIPENING

30 minutes

85° F.
2 gallons whole goat's milk

2 ounces mesophilic goat cheese starter culture

Warm 2 gallons of whole milk to 85° F. Add 2 ounces of mesophilic goat cheese starter culture and stir in thoroughly. Allow to ripen for 30 minutes.

RENNETING

1 hour 30 minutes

85° F.

1/4 rennet tablet (or 1 teaspoon liquid rennet)

Dissolve 1/4 rennet tablet (or 1 teaspoon liquid rennet) in 1/4 cup cool water. Add to the ripened milk and gently stir for several minutes. Allow to set for 60 minutes.

CUTTING THE CURD

1 hour 40 minutes	Cut the curd into 1/2-inch cubes. Allow the curd to set undisturbed for 10 minutes.

COOKING

3 hours	Raise the temperature of the curd 2 degrees F. every 5 minutes until the temperature reaches 98° F. Gently stir often. Allow the temperature to remain at 98° F. for 45 minutes. Continue to stir often (gently).

DRAINING

Drain the curds into a cheesecloth-lined colander.

MOLDING

3 hours 15 minutes	Line a 2-pound cheese mold with cheesecloth. Quickly place the curds into the mold. Cover with a follower and press the cheese at 20 pounds pressure for 15 minutes.
4 hours 15 minutes	Flip the cheese over in the mold and press at 30 pounds pressure for 1 hour.

Flip the cheese over and press at 50 pounds pressure for 12 hours.

SALTING

Remove the cheese from the press. Gently remove the cheesecloth. Rub salt on all surfaces of the cheese.

AGING

Place the cheese to age at 50° F. Rub salt on the cheese once a day for the next two days. Turn the cheese daily. When the surface of the cheese is dry, it may be waxed. The cheese should be turned daily. It can be eaten after aging for 1 month but improves in flavor if aged longer.

Whole Goat Milk Ricotta

This is an easy recipe for a whole milk goat ricotta which has a lot of flavor and gives a good yield of up to 2 pounds per gallon of milk. *Makes 4–8 ounces.*

ACIDIFYING AND COAGULATING

206° F.
1 gallon whole milk
1/4 cup vinegar

Warm 1 gallon of whole goat's milk to 206° F. Stir in 1/4 cup cider vinegar. The milk will rapidly coagulate.

DRAINING

Drain the curd into a cheesecloth-lined colander. Drain for a minute. Place the curds in a bowl. Mix 3 tablespoons melted butter and 1/2 teaspoon baking soda into the curd. Mix in thoroughly. Place the cheese in a container, cover, and refrigerate until ready for use. This cheese is excellent for cooking.

BACTERIAL AND MOLD-RIPENED CHEESES

Bacterial and mold-ripened cheeses owe much of their flavor and texture to the growth of a special mold or bacteria on the surface or on the surface and in the interior of the cheese. Most of these are soft cheeses. They all have the following basic steps in their production:

Ripening	**Molding and draining**
Renneting	**Salting**
Cutting	**Application of mold or bacteria**
Cooking	**Aging**

The cheeses included in this chapter are Camembert, Coulommiers, blue, Stilton, Gorgonzola, and brick.

Ripening

For most of the soft mold-ripened cheeses, a fairly large amount of cheese starter culture is added, and the ripening time can be rather lengthy. There is a goodly increase in acidity developed during ripening, and the recipes call for mesophilic starter.

Renneting

A comparatively small amount of rennet is used to set most mold-ripened cheeses. It may take up to an hour for the milk to coagulate, and the curd is slightly less firm than curd produced for hard cheeses.

Cutting the Curd

The curd is usually cut into 1/2-inch cubes, although sometimes the curds are ladled in thin slices directly into a cheese mold.

Cooking

The cooking temperatures are quite low, usually around 90° F. Some recipes omit this step.

BACTERIAL AND MOLD-RIPENED CHEESES

Variety	*Amt. Milk*	*Type of Milk*	*Coagulation*	*Culture*	*Amt. of Cheese*	*Use of Cheese*
Camembert	2 gal. whole	Cow's	Rennet 1/4 tsp.	Mesophilic 4 oz.	2 lbs. +	Table, cooking
Coulommiers English Style	1/2 gal. whole	Cow's	Rennet 3 drops	Mesophilic 1 oz.	3/4 lb.	Table
Herbed Coulommiers	1/2 gal. whole	Cow's	Rennet 3 drops	Mesophilic 1 oz.	3/4 lb.	Table
Coulommiers French Style	1/2 gal. whole	Cow's	Rennet 3 drops	Mesophilic 1 oz.	3/4 lb.	Table
Blue Cheese	2 gal. whole	Cow's, Goat's		Mesophilic 4 oz.	2 lb.	Table, salad, cooking
Stilton	2 gal. whole, 2 cups cream	Cow's	Rennet 1/4 tsp.	Mesophilic 2 oz.	2 lb.	Table
Gorgonzola	2 gal. whole	Cow's	Rennet 1/2 tsp.	Mesophilic 2 oz.	2 lb.	Table
Brick	2 gal. whole	Cow's	Rennet 1/4 tab., 3/4 tsp.	Mesophilic 2 oz.	2 lb.	Table

Molding and Draining

The curds are placed into cheese molds. Many of these are a traditional shape and size. The molds are usually placed on wooden reed cheese mats or plastic cheese mats, situated so that the whey will readily drain into the sink. The mats rest on a cheese board because the molds must be flipped over quite often and the cheese boards make this task easier.

So, with a mold resting on a cheese mat which is resting on a cheeseboard, you can fill the mold with curd. The mold is filled to the top. Whey immediately starts to drip out of the mold holes and through the reed cheese mat on the bottom of the mold. A second cheese mat is placed on top of the mold and a wooden cheese board is placed on the mat. These are unpressed cheeses, and the curd sinks under its own weight. After a period of draining, the cheese is flipped over. The cheese mat which is now on the top is gently peeled off the mold to allow the cheese to fall to the bottom of the mold. A fresh mat is put on top of the mold, followed by a cheese board, and the cheese continues to drain for 24 to 35 hours.

Salting

Once the cheese has properly drained in the mold and assumed its shape, it is taken from the mold and salted on all its surfaces. The salt will help retard the growth of unwanted organisms but will not interfere with the mold growth which you desire.

Application of Mold or Bacteria

Mold and bacterial cultures can be purchased in 10-gram packets of freeze-dried powder which will keep 2 years in the freezer.

White mold powder (*Penicillium candidum*) is used for aging Camembert and other related cheeses. Place 1 quart of cool water in an atomizer and add the packet of powder. Shake the solution several times and then lightly spray it over the surface of the cheese.

Blue mold powder (*Penicillium roqueforti*) is used in making the blue cheeses. This powder is usually added to the curd just before it is placed into the mold. The mold powder can be mixed with salt which must be added to the cheeses. This mixture is gently mixed throughout the curds, which are then placed in a cheese mold. Unlike Camembert cheese, in which the mold simply grows on the surface of the cheese, the blue mold grows throughout blue cheese.

The red bacteria is used to age brick cheese which is like a mild Limburger. Add the red bacteria powder to 1 quart of water in an atomizer, and lightly spray the cheese surface.

Aging

These cheeses need to be aged in a rather cool, moist setting in order to allow the mold or bacteria to grow. The temperature is usually at 45° F. and the relative humidity is near 95 percent which is very moist. To create this environment in the home is a challenge. An extra second-hand refrigerator can be very useful. A tray of water is placed on the bottom of the refrigerator for moisture and the setting is kept at 45° F. If an extra refrigerator will not fit into your budget, you can try to age these cheeses in your family refrigerator. It is best to have the cheese rest on a cheese mat in a closed container which also has a bowl of water in it for moisture.

These cheeses need to be aged on mats so that air can circulate under the cheese. The white mold-ripened cheese will develop a furry, thick, white coat within 14 days after which the cheeses should be wrapped in Saran Wrap and aged an additional 6 weeks.

On blue cheeses, the mold will grow within 10 days. At this time the cheese must be pierced with holes so that oxygen will get to the interior of the cheese and aid in mold growth.

On red bacterial-ripened cheeses, the bacteria will show up in 2 weeks as a reddish brown smear on the surface of the cheese.

WHITE MOLD-RIPENED CHEESES

Camembert Cheese

Camembert cheese originated in the village of the same name in France in 1791. It was discovered by Marie Fontaine and became one of the most prized cheeses in the world. It is a surface mold-ripened cheese and has the mold *Penicillium camembertii* or *Penicillium candidum* growing on its surface during curing. This mold develops the sharp taste of the cheese and aids in creating its runny consistency. The cheese is difficult to make because of the exacting conditions required during curing. It must be stored at 45° F. with a relative humidity of 95 percent. The cheese may be made with either goat's or cow's milk.

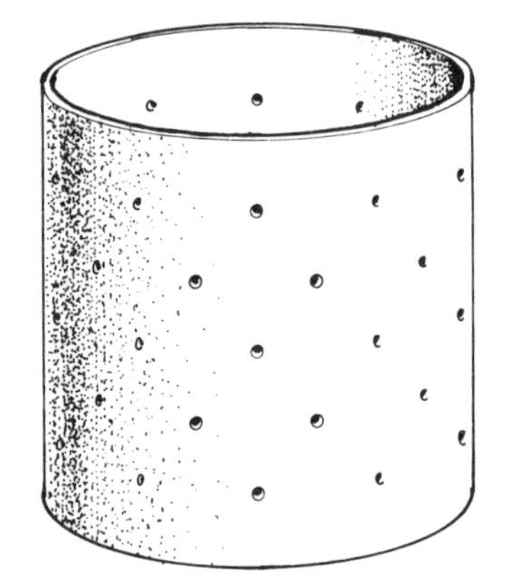

Camembert mold is distinctive in size and shape.

One-half gallon of milk will make enough curd to fill a standard Camembert mold. A Camembert mold is a cylinder open at both ends 4 3/8 inches in diameter and 4 5/8 inches in height. The mold has drainage holes along its sides. *Makes over 2 pounds.*

RIPENING

1 hour 30 minutes	**90° F.** **2 gallons whole milk**	**4 ounces mesophilic cheese starter culture**

Warm 2 gallons of whole milk to 90° F. and add 4 ounces of mesophilic cheese starter culture. Mix in thoroughly. Allow the milk to ripen for 1 1/2 hours.

RENNETING

2 hours 30 minutes — **1/4 teaspoon liquid rennet**

Add 1/4 teaspoon liquid rennet to 1/4 cup cool water and pour into the ripened milk. Mix in thoroughly yet gently for several minutes. Let the milk set for 60 minutes or until the curd gives a clean break.

CUTTING THE CURD

2 hours 45 minutes — **90° F.**

Cut the curd into 1/2-inch cubes. Gently stir the curds for 15 minutes.

MOLDING AND DRAINING

3 hours 10 minutes

Sterilize the molds in boiling water. You should also sterilize the wooden reed cheese mats and cheese boards. For 2 gallons of milk you will need four molds, eight mats, and eight cheese boards (6″ × 6″). Place the boards, mats and molds in a convenient spot to drain. Place a reed mat on a cheese board and a Camembert mold on the mat. Let the pot of curds set 15 minutes, then pour off the whey to the level of the curds. Ladle the curds gently into a Camembert mold until full. Place a reed mat on top of the mold and a cheese board on top of the mat. Fill the three remaining molds.

4 hours

Let the molds drain for 1 hour, then flip the molds over so that the top is on the bottom. Make sure that the cheese is not sticking to the top mat by gently peeling the mat from the top of the mold. You do not want to rip the surface of the cheese, which will happen if you lift the mat briskly.

9 hours

Turn the cheeses at hourly intervals for 5 hours or until the cheese is about 1 to 1 1/2 inches in height and has shrunk from the sides of the mold.

SALTING

9 hours 10 minutes

When the cheese has sufficiently drained, sprinkle coarse cheese salt on all surfaces of the cheese and let it rest for 10 minutes for the salt to dissolve.

APPLYING WHITE MOLD SPORES

Prepare the white mold by placing a quart of water in an atomizer. Empty the contents of the powdered mold into the atomizer. The powder will not dissolve so shake the atomizer to mix well. Lightly spray all surfaces of the cheese. The cheese should not appear wet.

AGING

Place the cheese on a plastic or wooden reed cheese mat and age at 45° F. and 95 percent relative humidity. Let the cheese set for 5 days or until the first whiskers of mold appear on the surface. When white mold appears, turn the cheese over and leave for 9 days. After 14 days of aging a profuse white mold will have developed. Wrap the cheese in cellophane (not plastic wrap as the cheese needs to breathe) and store at 45° F. for 4 to 6 weeks. The cheese is ready to eat when it has a runny consistency throughout when served at room temperature. That's about six weeks after wrapping.

If blue mold develops on the cheese, your curing room is too humid or you are allowing too much moisture to remain in the cheese during manufacture. Reduce the humidity in the curing room and thoroughly clean and disinfect all shelves to eliminate blue mold.

Coulommiers Cheese

Coulommiers cheese is a soft white mold-ripened cheese which is in the same family as Camembert. It is a cheese traditional to France and England. The French recipe is for a soft, runny, textured cheese as is Camembert. The English cheese is consumed fresh and is a soft, spreadable cheese which is not mold-ripened.

The traditional mold for Coulommiers is a two-piece affair consisting of two stainless steel hoops, one of which fits inside the other. The assembled hoops are 6 inches in height and 4 3/4 inches in diameter. They are filled with curd, and once it sinks below the level of

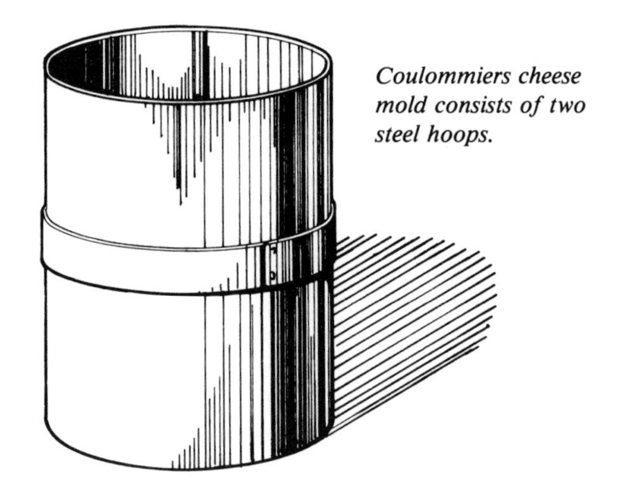

Coulommiers cheese mold consists of two steel hoops.

the top hoop, the top hoop is removed and the cheese is flipped over. Coulommiers means column in French, and that is what the mold looks like. If the cheese mold were not in two sections, the draining cheese would splatter when turned over in the tall mold. One Coulommiers mold will make a cheese from 1/2 gallon of whole milk. *Makes 3/4 pound.*

Coulommiers Cheese

English Style

RIPENING

20 minutes

90° F.
1/2 gallon whole milk
1 ounce mesophilic cheese starter culture

Warm 1/2 gallon of whole milk to 90° F. Add 1 ounce of mesophilic cheese starter culture and mix in thoroughly. Cover and leave to ripen for 20 minutes.

RENNETING

1 hour 5 minutes

90° F.
3 drops liquid rennet
Coulommiers mold
2 cheese mats
2 cheese boards

Dissolve 3 drops of liquid rennet in 2 tablespoons of cool water. Gently stir into the milk for several minutes. Cover and allow to set for 45 minutes.

In boiling water, sterilize one Coulommiers mold, two cheese mats, and two cheese boards. Place a cheese mat on a board and place the Coulommiers mold on the mat. This should be set in an area where the whey can freely drain.

MOLDING AND DRAINING

10 hours 5 minutes

Using a stainless steel ladle, skim thin layers of curd from the pot and gently lower them into the mold. Use one hand to steady the mold as you gently place the curd slice into the bottom of the mold. Continue to ladle thin slices of curd into the mold until it is filled. Be careful not to tip the mold or all the curd will rush out the bottom. The thinner you slice the curd, the faster the cheese will drain. Place a second cheese mat on top of the filled mold and place a cheese board on top of the mat. Let the mold drain for 6 to 9 hours in an area where the whey can drain and the room temperature will stay near 72° F.

Day 3

When the curd has sunk below the level of the top hoop, remove the hoop and place the cheese mat and board on top of the remaining hoop. Very carefully lift the mold and cheese boards between two hands and quickly flip the cheese over. Set the mold down and gently remove the top mat, making sure not to tear the cheese, which may be sticking to it. Clean the cheese mat and replace on top of the hoop. Replace the board. Flip the cheese in a similar fashion several times a day for the next two days.

SALTING

Allow the cheese to drain for up to 2 days, flipping the mold over several times a day. The cheese is done when it stands 1 to 1 1/2 inches in height and has pulled away from the sides of the mold. Salt it lightly on all surfaces. It may be eaten right away. It may be wrapped in cellophane and aged for 1 to 2 weeks.

Coulommiers Cheese

French Style

The French style Coulommiers cheese is made with the same recipe as the English style. When the cheese is removed from the Coulommiers mold it is lightly salted and *lightly sprayed with a solution of white mold powder.* The cheese is aged for 5 days at 45° F. with a relative humidity of 95 percent. After 5 days, small whiskers of white mold will be seen growing on the surface of the cheese. Turn the cheese over and age it for 9 more days, at which time the cheese will be covered with a thick growth of white mold. Wrap the cheese in cellophane and age it 4 to 6 weeks at 45° F. *Makes 3/4 pound.*

Herbed Coulommiers

English Style

As the curds are ladled into the Coulommiers mold, herbs may be sprinkled between the slices of curd. A tasty combination is to add a clove of finely diced garlic, a few sprigs of chopped chives, a pinch of freshly ground black pepper, and a dash of paprika. *Makes 3/4 pound.*

BLUE MOLD-RIPENED CHEESES

Few things in this life can rival the satisfaction of cutting into a wheel of your own homemade blue cheese.

Blue mold-ripened cheeses have been made since the time of the Romans. The term Roquefort first appeared in 1070 and was a blue-veined cheese made from ewe's milk. The blue mold is *Penicillium roqueforti,* and it is cultivated by growing it on rye bread. Blue cheeses are usually not pressed and take their shape as the curds sink under their own weight in a cheese mold. The cheeses are commonly pierced with needles to form air holes so that the blue mold, which requires oxygen, can grow in the interior.

Salt is used in fairly large amounts in the curing of these cheeses; the blue mold does not mind salt, and undesirable molds are unable to flourish in a salty environment.

Blue cheeses are not for beginners to make. In many ways a good homemade blue cheese is like one of your children, easy and fun to make, but difficult to bring to a proper maturity. These cheeses must be aged in very damp, cool places for several months. If the proper environment is not maintained, they can dry out and also develop undesirable molds on their surface.

The blue mold is available in 10-gram packets of freeze-dried powdered spores. Kept in an airtight container in the freezer, this spore powder will be viable for up to 18 months. The blue mold can be added directly to the milk or it can be sprinkled among the curds as they are placed in the cheese mold. The easiest method is to mix the spore powder with salt and sprinkle this on the curds.

The three recipes which follow are listed in order of difficulty: blue cheese, Stilton, and Gorgonzola.

Blue Cheese

This cheese is made with whole milk. If using cow's milk, the cheese will have a light yellow color with the blue mold visible throughout the interior. If using goat's milk, the cheese will be white. The flavor of this cheese is in large measure a result of the blue mold *Penicillium roqueforti* which grows on the surface and in the interior of the cheese. *Makes 2 pounds.*

RIPENING

1 hour	**2 gallons whole milk** **90° F. (cow)** **86° F. (goat)**	**4 ounces mesophilic cheese starter culture**

To 2 gallons of whole cow's milk at 90°

F. (86° for goat's milk) add 4 ounces mesophilic cheese starter culture. Mix thoroughly. Let set undisturbed at a constant temperature for 60 minutes.

RENNETING

1 hour 45 minutes — **1 teaspoon liquid rennet (or 1/4 rennet tablet)** — **90° F. (cow)** **86° F. (goat)**

Add 1 teaspoon liquid rennet (or 1/4 rennet tablet) to 1/4 cup cool water. Add the rennet solution to the milk and gently stir for 1 minute. Let the milk set for 45 minutes or until a clean break is evident.

CUTTING

1 hour 50 minutes — **90° F. (cow)** — **86° F. (goat)**

Cut the curd into 1/2-inch cubes. Allow them to set for 5 minutes to firm up before stirring. (If using goat's milk, let the curds set for 10 minutes).

COOKING

2 hours 50 minutes — **90° F. (cow)** — **86° F. (goat)**

Gently stir the curds every 5 minutes to keep them from matting together. Do this for 60 minutes.

DRAINING

3 hours — **90° F. (cow)** — **86° F. (goat)**

Let the curd set undisturbed for 5 minutes. Pour off the whey. Place the curd in a colander and allow to drain for 5 minutes. Place the curds back into the pot and mix them gently by hand so that the curd pieces are not matted together.

SALTING

3 hours 5 minutes — **Salt** — **Blue Mold Powder**

To 2 tablespoons coarse flake salt add 1/4 teaspoon blue mold spore powder. Mix well. Sprinkle half of the salt-mold mixture over the curds and gently mix it in. Sprinkle the remaining half and gently mix it in. Let the curds rest for 5 minutes.

MOLDING

3 hours 10 minutes — **2-pound cheese mold** — **2 cheese boards** — **2 cheese mats**

Place a 2-pound cheese mold on a cheese mat which is resting on a cheese board. Place it where the whey can drain off. Fill the mold with the curds. Cover with a cheese mat and cheese board.

7 hours 10 minutes Turn the mold over every 15 minutes for the first 2 hours, and then once an hour for the next 2 hours.

Day 2 Leave to drain overnight.

SALTING

Day 3

Coarse flake salt

Take the cheese from the mold and sprinkle a coarse flake salt on all its surfaces. Shake off excess salt. Place the cheese where it will stay at 60° F. with a relative humidity of 85 percent. Many basements will meet this requirement. Rest the cheese on a cheese mat or board.

Day 6

Salt

Turn the cheese over each day and resalt it, shaking off excess salt for the next 3 days.

CURING

10 days

Ice pick or Knitting needle

Using an ice pick or knitting needle with a 1/16-inch diameter, put forty holes through the top of the cheese from top to bottom. Place the cheese where the temperature is 50° F. and the relative humidity is 95 percent. An extra refrigerator with a pan of water on the bottom works very well. It is best to store the cheese on its side in a wooden cradle. Every 4 days give the cheese a quarter turn on its side to keep it from becoming misshapen.

The blue mold will usually be seen growing after 10 days.

Use ice pick or knitting needle for piercing holes.

1 month After 30 days the surface of the cheese will be covered with blue mold and a reddish-brown smear. This should be gently scraped off with a long-bladed knife.

The cheese should be scraped of mold and smear every 20 to 30 days.

3 months After 90 days of curing, the cheese should be scraped and wrapped in Saran Wrap or tin foil. The cheese should be stored in a refrigerator at 34°–38° F. for 2 more months. It should be turned weekly.

6 months The cheese is ready to eat after 6 months. If a milder cheese is desired, you may wish to sample a wheel after 3 months.

Stilton

Stilton is an English cheese which was first mentioned in the early part of the eighteenth century. It is made from milk enriched with cream. It has a subtle taste of Cheddar with the sharp taste of blue cheese. *Makes 2 pounds.*

RIPENING

30 minutes **86°**
2 gallons whole milk
2 cups cream
2 ounces mesophilic starter culture

To 2 gallons of whole milk add 2 cups cream. Mix thoroughly. Warm the milk to 86° F. Add 2 ounces mesophilic cheese starter culture. Let ripen for 30 minutes.

RENNETING

2 hours **86° F.** **1/4 teaspoon liquid rennet**

Add 1/4 teaspoon liquid rennet diluted in 1/4 cup water to the milk. Gently stir, using an up-and-down motion, for 1 minute. Top-stir for several minutes. Let set for 90 minutes.

CUTTING

2 hours 10 minutes Cut the curds into 1/2-inch cubes. Let stand for 10 minutes.

DRAINING

3 hours 40 minutes **86° F.**

With a slotted spoon transfer the curds into a cheesecloth-lined colander resting in a bowl. When the curds are in the cheesecloth-lined colander they should be surrounded by whey in the bowl. Let the curds rest in the whey for 90 minutes.

4 hours 10 minutes Hang the bag of curd to drain for 30 minutes.

2 cheese boards **8-pound weight**

When the curd bag has stopped dripping, place it on a board where it can drain. Place a board on top of the curd bag and place 8 pounds of weight on the board. Press the cheese overnight in a kitchen which has a temperature of 68°–70° F.

MILLING AND SALTING

Salt **Blue mold powder**

Remove the curd from the bag. Break into 1-inch pieces and place in a bowl.

To 2 tablespoons of coarse flake salt add 1/4 teaspoon blue mold powder. Mix thoroughly. Sprinkle the mold-salt mixture over the curds using half of the mixture at a time. Mix the salt into the curds thoroughly but gently.

MOLDING AND DRAINING

6 days **2-pound cheese mold** **2 cheese mats**

Place the curds into a 2-pound cheese mold resting on a cheese mat and cheese board. Cover the top of the mold with a cheese mat and cheese board. Flip over the cheese mold every 15 minutes for 2 hours. The kitchen should be at 70° F. Leave to set overnight. Flip the cheese several times a day for the next 4 days.

4 months 6 days **Ice pick or Knitting needle**

Using a 1/16-inch diameter ice pick or knitting needle, put twenty-five holes from the top to the bottom of the cheese. Place the cheese on a cheese mat in a 50°–55° F. chamber with a relative humidity of 90 percent. Turn the cheese several times a week. Scrape the cheese of mold and slime once a week. The cheese should be cured for 4 months before being consumed.

Gorgonzola

This cheese originated in Italy and is named after a village outside of Milan. This is a rich, creamy cheese with a flavor enhanced by the growth of blue mold. *Makes 2 pounds.*

RIPENING

30 minutes **86° F.** **1 gallon whole milk** **2 ounces mesophilic cheese starter culture**

Warm 1 gallon of whole milk to 86° F. Add 2 ounces mesophilic cheese starter. Allow to ripen for 30 minutes.

RENNETING

1 hour 15 minutes — **86° F.** — **1/2 teaspoon liquid rennet (1/4 rennet tablet)**

Add 1/2 teaspoon liquid rennet to 1/4 cup cool water. Add to the milk. Gently stir for 1 minute. Let the milk set for 45 minutes.

CUTTING

1 hour 25 minutes — Cut the curds into 1/2-inch cubes. Let set for 10 minutes.

DRAINING

1 day — Pour the curds into a cheesecloth-lined colander. Tie the four corners of the cheesecloth and hang to drain overnight in a kitchen with a temperature of 68°-70° F.

Make a second batch of curds from 1 gallon of milk as was done the previous evening. Drain these curds in a cheesecloth bag for 1 hour.

MILLING

Cut the drained curds from the previous evening into 1-inch cubes. Place in a bowl. Cut the drained curds from the morning into 1-inch cubes and place in another bowl.

SALTING

Salt — **Blue mold powder**

To 4 tablespoons coarse flake salt add 1/4 teaspoon blue mold powder. Mix in thoroughly. Sprinkle half of the salt mold mixture in the first bowl of curd and half of the mixture in the second bowl. Mix gently but thoroughly.

MOLDING

2-pound cheese mold — **2 cheese boards** — **2 cheese mats**

Place the curds in a 2-pound cheese mold which is resting on a cheese mat and board. Place the morning curd on the bottom, outside, and top of the mold. Place the evening curd mostly in the center of mold. Flip the mold over every 15 minutes for the first 2 hours. Place the mold in a room where the temperature is 55°-60° F. Flip the cheese over several times a day for the next 3 days.

SALT

9 days Remove the cheese from the mold and sprinkle coarse salt over all surfaces of the cheese. Shake off excess salt. The cheese should be cured at 55° F. with a relative humidity of 85 percent. Each day for the next 4 days the cheese should be rubbed with a coarse salt.

3 months **Ice pick or Knitting needle**

Using an ice pick or knitting needle with a 1/16-inch diameter, put twenty-five holes into the cheese going from top to bottom.

After 30 days of curing, the cheese should be placed where the temperature is 50° F. and the relative humidity is 85 percent. A second refrigerator with a pan of water on the bottom works well. Leave the cheese to mature for 60 days longer. Scrape the cheese clean of all mold and smear, with a long-bladed knife every several weeks.

Four months after being produced, the cheese is ready for eating. It can be aged for several more months if desired.

Red Bacterial-Ripened Brick Cheese

Brick cheese was developed in the United States. Its characteristic shape is rectangular, measuring 10 by 5 by 6 inches. This cheese can be made in the traditional rectangular mold or any mold that will allow adequate drainage. Brick is a bacterial-ripened cheese. A reddish-brown bacteria (*Bacteria linens*) grows on the surface and gives the cheese its characteristic flavor. Since the flavor of the cheese is in large measure a result of the bacterial growth, the cheese cannot be thicker than those dimensions or the bacterial enzymes will not penetrate the interior of the cheese. Brick has a mildly sharp flavor and is affectionately known as "the married man's Limburger." *Makes 2 pounds.*

RIPENING

10 minutes **86° F.** **2 gallons whole milk** **2 ounces mesophilic cheese starter culture**

Warm 2 gallons of whole milk to 86° F. Add 2 ounces of mesophilic cheese starter culture and mix in thoroughly. Allow to ripen for 10 minutes.

RENNETING

40 minutes **86° F.** **1/4 rennet tablet (or 3/4 tsp. liquid rennet)**

If cheese coloring is desired, it should be added before the rennet. Dissolve 1/4 rennet tablet (or 3/4 teaspon liquid rennet) in 1/4 cup cool water. Stir the rennet into the milk and gently stir for several minutes. Allow the milk to set for 30 minutes or until the curd gives a clean break.

CUTTING THE CURD

50 minutes **86° F.**

Cut the curd into 1/4-inch cubes.

COOKING

1 hour **86° F.**

Stir the curd gently for 10 minutes. During cooking, the curds should be stirred to keep from matting.

1 hour 5 minutes **86° F.**

Raise the temperature of the curd 1 degree F. in 5 minutes.

1 hour 10 minutes **87° F.**

Raise the temperature 2 degrees in 5 minutes.

1 hour 15 minutes **89° F.**

Raise the temperature 3 degrees in 5 minutes.

1 hour 20 minutes **92° F.**

Raise the temperature 4 degrees in 5 minutes.

DRAINING AND WASHING

1 hour 50 minutes Allow the curds to set undisturbed for 5 minutes. Drain the whey down to the level of the curds. Add enough tap water at 96° F. to the curds to replace the whey that was drained off. Stir the curds for 20 minutes and keep at 96° F.

Let the curd set for 5 minutes undisturbed, then drain off the diluted whey.

MOLDING AND PRESSING

2 hours Ladle the curds into a 2-pound cheese mold, resting on a cheese mat and cheese board which can drain. After 10 minutes, flip the mold upside down.

2 hours 30 minutes Turn over again after 30 minutes.

7 hours 30 minutes Add a follower to the mold and press with a 5-pound weight. Turn the cheese several times during the next 5 hours, keeping cheese under that pressure.

SALTING

13 hours 45 minutes Remove the cheese from the mold. Place in a brine solution (2 pounds of salt in 1 gallon water) for 6 hours.

Remove the cheese from the brine. Dry with a paper towel. Make up a solution of *Bacteria linens* by adding 1 teaspoon of *Bacterial linens* powder to 1 quart of water in an atomizer. Lightly spray all surfaces of the cheeses.

AGING

2 weeks The cheese needs to be stored at 60° F. with a relative humidity of 90 percent for 14 days. Each day the cheese should be gently washed with a salted water (1/2 pound salt to 1 gallon water). Wet the palm of one hand with the salted water and dampen all surfaces of the cheese.

After 2 weeks the cheese will have a reddish-brown color due to the growth of *Bacteria linens.* Rinse the cheese in cool water. Gently dry with a paper towel. Wax the cheese and store it at 45° F. for 6 to 10 weeks. Turn it several times a week.

QUICK CHEESES

DIRECT-SET CULTURES

In the late 1980s, a new technology of home starter cultures, known as *direct-set cultures,* was introduced. These cultures are available in freeze-dried packets and can be added directly to milk for cheesemaking. There are now direct-set cultures available for making almost all cheese, both hard and soft. These cultures save a great deal of time in home cheesemaking and are very easy to use. There are also direct-set cultures available for making yogurt, buttermilk, and sour cream.

Fromage Blanc

Fromage blanc is a fresh, easy-to-make cheese. Of French origin, its name simply means "white cheese." It makes an excellent cheese spread and can have herbs and spices added to it. It can also be used by itself as a substitute for cream cheese or ricotta in cooking.

It has the consistency of a cream cheese with a fraction of the calories and cholesterol. It can be made with either whole or skim milk. Fromage blanc is made with a direct-set culture. *Makes up to 2½ pounds.*

ACIDIFYING AND COAGULATION

12 hours (or overnight)	1 gallon milk (whole or skim	1 packet of direct-set fromage blanc starter

Heat one gallon of milk to 180° F. and then cool to 72° F. Add 1 packet of direct-set fromage blanc starter to the milk and stir in. Cover and let set at 72° F. for 12 hours (or overnight).

DRAINING

6 to 12 hours — **cheesecloth** · **colander**

Line a colander with a layer of fine cheesecloth. Ladle the coagulated fromage blanc curd into the colander. Allow to drain for 6 to 12 hours or until it reaches the desired consistency. A shorter draining time produces a cheese spread. A longer draining time produces a cream cheese–type consistency. Make sure the room temperature is close to 72° F. during draining. After draining, store in a covered container and refrigerate.

Mascarpone

Mascarpone is a soft Italian cheese that is made from cream and used in cooking or in desserts such as Italian pastries. It may also be served by itself on slices of sweet bread. *Makes 10 to 14 ounces.*

ACIDIFYING AND COAGULATION

1 quart light cream · **¼ teaspoon tartaric acid***

In a double boiler, heat 1 quart of light cream (do not use heavy cream) to 180° F.

Add tartaric acid to the hot cream, and stir for several minutes. The cream should slowly thicken into a custard-like consistency with tiny flecks of curd noticeable. If the cream does not coagulate, add a speck more of tartaric acid and stir an additional 5 minutes. Be careful not to add too much tartaric acid, or a grainy texture will result.

DRAINING

12 hours (or overnight) — **cheesecloth** · **colander** · **bowl**

Line a stainless steel colander with a double layer of fine cheesecloth. Pour the curd into the colander and drain for one hour. Place the colander in a bowl and drain in the refrigerator for 12 hours (or overnight).

Place the finished cheese in a covered container and refrigerate. This cheese will keep for up to two weeks.

**Tartaric acid is a natural vegetable acid derived from the seed of the tamarind tree, which is found throughout the Caribbean.*

GLOSSARY

ACID CURD
The custard-like state that milk is brought to when a high level of acidity is created. The acidity is produced by the activity of starter culture bacteria, and it precipitates the milk protein into a solid curd.

ACIDITY
The amount of acidity (sourness) in the milk. Acidity is an important element in cheesemaking and it is produced by cheese starter culture bacteria.

AGING
A step in cheesemaking in which the cheese is stored at a particular temperature and relative humidity for a specified amount of time in order to develop its distinct flavor.

ALBUMINOUS PROTEIN
Protein in milk which cannot be precipitated out by the addition of rennet. Albuminous protein remains in the whey and is precipitated by high temperatures to make ricotta.

ANNATTO
A natural vegetable extract which is used to color cheese.

BACTERIA
Microscopic unicellular organisms found almost everywhere. Lactic acid-producing bacteria are helpful and necessary for the making of quality hard cheeses.

BACTERIA LINENS
A red bacteria which is encouraged to grow on the surfaces of cheeses like brick or Limburger to produce a sharp flavor.

BACTERIAL-RIPENED CHEESE
A cheese upon whose surface bacterial growth is encouraged to develop in order to produce a distinct flavor. Brick and Limburger are examples of bacterial-ripened cheeses.

BUTTERFAT
The fat portion (cream) in milk. Butterfat can vary from 2.5 to 5.5 percent of the total weight of milk.

CHEESE BOARD
A board measuring 6 inches square and 1 inch thick of maple or birch, often used to aid in the draining of soft cheeses such as Camembert. Larger cheese boards are often used to hold aging cheeses.

CHEESECLOTH

A coarse to finely woven cotton cloth used to drain curds, line cheese molds, and perform a host of other cheesemaking functions.

CHEESE COLOR

A coloring added to the milk prior to renneting which will impart various shades of yellow to the cheese. Most coloring is a derivative of the annatto tree.

CHEESE MAT

A wood reed cheese mat often used to aid in the drainage of soft cheeses such as Coulommiers or Camembert.

CHEESE SALT

A coarse flake salt. Salt not iodized is the most desirable type to use in cheesemaking.

CHEESE STARTER CULTURE

A bacterial culture added to milk as the first step in making many cheeses. The bacteria produce an acid during their life cycle in the milk. There are two categories of starter culture: mesophilic and thermophilic.

CHEESE WAX

A pliable wax with a low melting point which produces an airtight seal which will not crack. Most hard cheeses are waxed.

CLEAN BREAK

The condition of the curd when it is ready for cutting. A finger or thermometer inserted into the curd at a 45° angle will separate the curd firmly and cleanly if the curd has reached that condition.

COOKING

A step in cheesemaking during which the cut curd is warmed to expel more whey.

COULOMMIERS MOLD

A two-piece stainless steel mold consisting of two hoops, one resting inside the other. The mold is used for making Coulommiers cheese.

CURD

The solid custard-like state of milk achieved by the addition of rennet. The curd contains most of the milk protein and fat.

CUTTING THE CURD

A step in cheesemaking in which the curd is cut into equal-sized pieces.

DAIRY THERMOMETER

A thermometer which ranges from 0° F. to 212° F. and can be used to measure the temperature of milk during cheesemaking.

DRAINING

A step in cheesemaking in which the whey is separated from the curd by pouring the pot of curds and whey into a cheesecloth-lined colander.

DRIP TRAY

A tray which is placed under a mold during the pressing of a cheese. The drip tray allows the whey to drain into a sink or container.

HOMOGENIZATION
A mechanical breaking up of the fat globules in milk so that the cream will no longer rise in the milk.

LACTIC ACID
Acid created in milk during cheesemaking. Cheese starter culture bacteria consume the milk sugar (lactose) and produce lactic acid as a byproduct.

LACTOSE
The sugar naturally present in milk. Lactose can constitute up to 5 percent of the total weight of milk.

MESOPHILIC CHEESE STARTER CULTURE
A blend of lactic acid-producing bacteria which is used to produce cheeses when the cooking temperature is 102° F. or lower.

MILLING
A step in cheesemaking during which the curd is broken into smaller pieces before being placed in a cheese press.

MOLDING
A step in cheesemaking during which the curd is placed in a cheese mold. The cheese mold will help produce the final shape of the cheese and aids in drainage.

MOLD-RIPENED CHEESE
A cheese upon whose surface (and/or interior) a mold is encouraged to grow. Two types of mold are most common in cheesemaking. They are blue mold for the blue cheeses and white mold for Camembert and related cheeses.

PASTEURIZATION
The heating of milk to 145° F. for 30 minutes. This destroys pathogenic organisms which may be harmful to man.

PENICILLIUM CANDIDUM
A white mold which is encouraged to grow on the surface of a number of soft mold-ripened cheeses including Camembert.

PENICILLIUM ROQUEFORTI
A blue mold which is encouraged to grow on the surface and in the interior of a variety of blue cheeses.

PRESSING
A step in cheesemaking during which the curds are placed in a cheesecloth-lined mold and placed under pressure to remove more whey.

PROPER BREAK
A term used during the making of Swiss cheese. To make certain the curds are properly cooked, a handful is wadded into a ball. If the ball can be easily broken back into the individual curd particles, this is called a proper break.

RAW MILK
Milk which is taken fresh from the animal and has not been pasteurized.

RED BACTERIA
A special bacterial growth (*Bacteria linens*) which is encouraged to grow on cheeses such as brick and Limburger.

REDRESSING

The changing of cheesecloth on a draining or pressed cheese. This helps keep the cheesecloth from sticking to the cheese.

RENNET (ANIMAL)

Rennet is derived from the fourth stomach of a milk-fed calf. It contains the enzyme renin which has the ability to coagulate milk. Animal rennet is available in tablet or liquid form.

RENNETING

A step in cheesemaking in which rennet is added to milk in order to bring about coagulation.

RIPENING

A step in cheesemaking in which the milk is allowed to undergo an increase in acidity due to the activity of cheese starter culture bacteria.

SALTING

A step in cheesemaking in which coarse flake salt is added to the curds before molding or to the surface of the finished cheese.

SOFT CHEESE

A cheese which is not pressed, contains a high moisture content, and is aged for a comparatively short period of time.

THERMOPHILIC CHEESE STARTER CULTURE

A bacterial starter culture which is used for the making of cheeses which have a high cooking temperature. Recipes for Italian cheeses and Swiss cheese call for a thermophilic culture.

TOP-STIRRING

The stirring of the top 1/4 inch of non-homogenized milk during cheesemaking in order to keep the cream from rising immediately after rennet has been added to the milk.

WHEY

The liquid portion of milk which develops after coagulation of the milk protein. Whey contains water, milk sugar, albuminous proteins, and minerals.

WHITE MOLD

A white mold (*Penicillium candidum*) which is encouraged to grow on a number of soft cheeses in order to develop a pungent flavor. Camembert is perhaps the most famous of these cheeses.

TROUBLE-SHOOTING CHART

Problems	*Possible Causes*	*Possible Solutions*
Cheese tastes very bitter.	Poor hygiene in handling the milk and/or cheesemaking utensils.	Keep the milk in a cold sanitary environment until ready for cheesemaking. Keep all utensils absolutely clean and free from milkstone. (Milkstone is a milk residue which is deposited over a period of time on the surface of utensils. It can be removed by using a dairy acid-type cleaner.) Sterilize all utensils. If using raw milk, and cheeses are bitter, you should pasteurize the milk prior to cheesemaking.
	Excessive rennet may have been used.	Reduce the amount of rennet used.
	Excessive acidity may be developing during the cheesemaking process.	Take steps to reduce acidity. See pages 18, 21, 50, and 56.
	Too little salt may have been added to the curd after milling.	Increase the amount of salt added to the curds at salting.
Cheese tastes quite sour and acidy.	The cheese contains too much moisture.	Take steps to reduce the moisture content during cheesemaking. See page 55.
	The cheesemaking process developed too much acidity.	Take steps to reduce the acid production. See pages 18 and 21.
The cheese has little to no flavor.	The cheese was not aged long enough.	Age the cheese the proper amount of time.
	Insufficient acidity was produced during cheesemaking.	Take steps to increase the acidity during cheesemaking. See page 50 and 56.
The milk does not coagulate into a solid curd.	Too little rennet was used.	Increase the amount of rennet used.
	Poor quality rennet was used.	Use a high-quality rennet which has been stored in an appropriate manner.
	Rennet activity was destroyed by mixing with very warm water when diluting.	Dilute rennet in cool water.

Problems	*Possible Causes*	*Possible Solutions*
The milk does not coagulate into a solid curd.	Rennet was mixed in same container as was cheese color.	Do not contaminate rennet with cheese coloring.
	Dairy thermometer is inaccurate and the setting temperature was actually too low.	Check accuracy of dairy thermometer.
	Milk contains colostrum.	Do not use milk which contains colostrum.
After adding rennet the milk almost instantly coagulates into a curd of tiny grains while the rennet is still being stirred into the milk.	Excessive acidity in milk.	Take steps to reduce acidity. Milk should not start to coagulate until about five minutes after adding rennet. To reduce acidity see pages 18, 21, 50, and 56.
The finished cheese is excessively dry.	May be caused by insufficient rennet.	Add more rennet.
	May be caused by cutting the curd into particles which are too small.	Cut the curd into larger pieces.
	Curds may have been cooked to an excessive temperature.	Lower the cooking temperature.
	Curds may have been overly agitated.	Treat curds gently.
Mold growth occurs on the surface of the air-drying cheese or a waxed cheese.	May be due to unclean aging conditions and/or too high a humidity in the aging room.	Clean all cheese aging shelves thoroughly. Lower the humidity of the cheese storage room.
The cheese is quite difficult to remove from the mold after pressing.	It may be that coliform bacteria and/or wild yeast have contaminated the milk and the curd. They have produced gas which has swelled the cheese during pressing. This production of gas makes it difficult to remove the cheese from the mold after pressing.	Pay strict attention to hygiene. Clean all utensils scrupulously. Sterilize all utensils in boiling water or in a sterilizing solution. Keep the milk clean and cold prior to cheesemaking. If using raw milk start to pasteurize the milk.
The cheese, when cut open, is filled with tiny holes, giving the cheese the appearance of a sponge.	May be coliform bacterial and/or wild yeast contamination. Such a contamination will be noted during the cooking process. The curds will have an unusual odor very similar to the smell of bread dough.	Pay strict attention to hygiene.

Problems	*Possible Causes*	*Possible Solutions*
The cheesecloth removes with great difficulty from the cheese after pressing. Pieces of the cheese may actually rip off when the cheesecloth is removed.	Coliform bacterial and/or wild yeast contamination.	Pay strict attention to hygiene.
	The cheese was not redressed in a fresh cheesecloth when needed. This is particularly true for cheeses made with a thermophilic culture.	Redress the cheese promptly as the recipe states.
Cheese becomes oily when air-drying.	Cheese is being air-dryed at too high a room temperature.	Remove the cheese to a cooler room. The temperature should not exceed 65° F.
	The curd was stirred too vigorously.	Handle the curd more gently.
	The curd was heated to too high a temperature.	Lower the cooking temperature.
Moisture spots are observed on the surface of the aging cheese beneath the wax. The wet spots can begin to rot and ruin the cheese.	Cheese was not turned often enough.	Turn the cheese at least daily when it first starts to age.
	Cheese contains excessive moisture.	Take steps to reduce the moisture content of the cheese. See page 55.
Lack of acidity during cheesemaking.	The starter culture is not working.	Antibiotics may be present in the milk. Do not use milk from animals receiving antibiotics or any other medication.
		The starter may be contaminated. Use a new starter.
		The presence of cleaning agents residue on utensils particularly chlorine. Rinse all utensils thoroughly.
Excessive acidity in cheesemaking.	Milk which has been improperly stored prior to cheesemaking or pasteurization.	Cool milk immediately after milking to below 38° F. Store milk at this temperature or below until ready for cheesemaking.

Problem	*Possible Causes*	*Possible Solutions*
Excessive acidity in cheesemaking.	Too much starter is being added.	Reduce the amount of starter used.
	Ripening period is too long.	Reduce the ripening time and add rennet sooner.
	Excessive moisture present in cheese.	Take steps to reduce moisture in cheese. See page 55.
Excessive moisture in cheese.	Inadequate acid development during cheesemaking.	Increase acidity during cheesemaking. See pages 50 and 56.
	Using milk which has too high a fat content.	The butterfat content of milk should not be much higher than 4.5%. If fat content is higher than this and excessive moisture is a problem remove some cream prior to cheesemaking.
	May be caused by heating the curds too rapidly. Too fast an increase in cooking temperature develops a membrane around the curd particles which does not allow the moisture to leave the curd particles.	Do not heat the curd faster than 2° F. every 5 minutes.
	May be caused by the retention of too much whey in the curd.	Cut the curd into smaller sizes.
	May be caused by heating the curd to too low a temperature during cooking.	Heat the curd at cooking to a somewhat higher temperature.

Mailorder Sources Of Cheesemaking Supplies

Caprine Supply
33001 West 83rd
P.O. Box Y
Desoto, KS 66018
Limited supply of starter cultures, kits, molds, presses, and miscellaneous equipment. Specializing in dairy goat supplies.

Cumberland General Store
Route 3, Box 81
Crossville, TN 38855
Starter cultures, presses, boxes, cutters, books, and miscellaneous tools.

Lehman Hardware
P.O. Box 41
Kidron, OH 44636
Starter cultures, kits, dairy thermometers, cheese presses, cheesecloth, butter churns, butter molds, and cheese colors. Catalog $2.

New England Cheesemaking Supply Co.
85 Main Street
Ashfield, MA 01330
Starter cultures (including direct-set starters), rennet, wax, molds, presses, kits, books, and other supplies. Instructional workshops.

Newsletters

The *Cheesemakers' Journal* is a bimonthly publication with colorful articles on home cheesemakers across the United States. Each issue includes a large recipe section.

Cheesemakers' Journal
85 Main Street
Ashfield, MA 01330

INDEX

Acid coagulation, *see* Coagulation
Acidity, 16
  control of, 21, 50–51, 69
Aging cheese:
  bacterial and mold-ripened types, 103–104
  hard cheeses, 60–61
Anatto, 22

Bacteria, red, 103, 115
Bacterial-ripened cheese, 101
  brick cheese, 115–17
  steps in making of, 101–104
Bacterial starter culture, *see* Starter culture
Bag cheeses, 29
  Bondon, 38–39
  buttermilk, 33–34
  Gervais, 37–38
  kefir, 35–36
  lactic, 30–31, *illus.* 31–32
  lemon, 36
  Neufchatel, 36–37
  queso blanco, 29–30
  yogurt, 34–35
Beginners, cheeses recommended for, 25, 119–120
Blue cheese, 109–12
Blue mold powder, 103, 109
Blue mold-ripened cheeses:
  blue, 109–12
  Gorgonzola, 113–15
  Stilton, 112–13
Board (for draining), 9 and *illus.,* 102
Bondon cheese, 38–39
Brick cheese, 115–17
Brine solution, 59–60
Butterfat, 11
Buttermilk, 19
Buttermilk cheese:
  dry, 33–34
  moist, 33
  "real," 34

Camembert cheese, 104–106
Caraway-flavored cheese:
  Cheddar, 66
  Gouda, 71
  Swiss, 76
Casein, 11
Cheddar cheese, 61
  caraway-flavored, 66
  cheddaring method, 61–63
  of goat's milk, 97–98
  jalapeno-flavored, 66
  sage-flavored, 65–66
  see also Derby cheese; Leicester cheese
Cheesecloth, 7
  draining curds in, 56 and *illus., illus.* 57
  ladling curds into, 15 and *illus.*
  removing from cheese, 59 and *illus.*
  see also Molding cheese; Redressing cheese
Cleanliness, 5
Coagulation, 14
  acid method: with acid substance, 14–15 and *illus.;* with bacterial starter culture, 15–18, *illus.* 17
  failure of (and solutions), 18–19
  rennet method, 20–21 and *illus.*
  and temperature, 21
Colby cheese, 71–73
Coloring, 22
  for hard cheese, 51
  from herbs, 24

Cooking curds, 55–56 and *illus.*
of bacterial and mold-ripened cheeses, 101
Cottage cheese, 42–43
of goat's milk, 45
large curd, 44–45
small curd, 43–44
Coulommiers cheese, 106–107
English style, 107–108
French style, 108
herbed, 108
Cream cheese, 39
cooked curd method, 40–41
French style, 42
Swiss style, 41
uncooked curd method, 39–40
Curd knife, *illus.* 6, 7
for cutting curd, 52–53 and *illus.*, *illus.* 54
Curd(s), 15, 52
components of, *table* 12
cooking, 55–56 and *illus.*, 101
cutting, 52–53 and *illus.*, *illus.* 54, 55, *illus.* 79, 101
draining, 56 and *illus.*, *illus.* 57, 102–103
of goat's milk, 93
ladling into cheesecloth, 15 and *illus.*
milling, 56, *illus.* 57
molding, 57–58 and *illus.*, 102
stirring (while cooking), 55 and *illus.*
Cutting curds:
for bacterial and mold-ripened cheeses, 101
for hard cheeses, 52–53 and *illus.*, *illus.* 54, 55
for Mozzarella cheese, *illus.* 79

Dairy thermometer, 6 and *illus.*, 50
Derby cheese, 61, 66–67
Direct-set cultures, 119
Draining curds:
of bacterial and mold-ripened cheeses, 102–103
of hard cheeses, 56 and *illus.*,
Drying cheese, 59
Dyes and dyeing, 22
and hard cheese, 51
herbs for, 24

Emmenthaler cheese, 76–77
starter culture for, 73, 76
Equipment, 5–9 and *illus.*
board (for draining), 9 and *illus.*, 102
curd knife, *illus.* 6, 7
curd stirrer, automatic, *illus.* 55
mat (for draining), 9 and *illus.*, 102, 103
molds, 7 and *illus.*, 8, 57–58 and *illus.*, 104, 106 and *illus.*
pots, 6–7 and *illus.*
presses, 8 and *illus.*, *illus.* 9
sterilizing of, 5–6
suppliers of (list), 129
thermometer, 6 and *illus.*, 50

Factories, development of, 2–3
Farmer's cheese, *see* Cottage cheese
Feta cheese, 96–97
Follower, 58 and *illus.*
Freezing starter culture, 17–18 and *illus.*
French cream cheese, 42
Fromage Blanc, 119–120

Gervais cheese, 37–38
Gjetost cheese, 89
Glossary, 121–24
Goat's milk, 12, 93
pasteurizing (method), 13–14 and *illus.*
Goat's milk cheeses, 93
Cheddar, 97–98
cottage cheese, 45
feta, 96–97
Ricotta, 98–99
Saint Maure, 96
soft, 93–95, *illus.* 95; herbed, 95
varieties, *table* 94
Gorgonzola cheese, 113–15
Gouda cheese, 69–71
caraway-flavored, 71
with hot peppers, 71

Hard cheeses, 47
steps in making of, 47, 50–61 and *illus.*
varieties, *table* 48–49
for specific recipes see Cheddar cheese; Italian cheeses; Swiss cheeses; Washed curd cheeses
Heating curds, *see* Cooking curds
Herbed cheeses, 24
Coulommiers, 108
soft goat cheese, 95
see also Caraway-flavored cheese; Sage Cheddar cheese
Herbs, 24
History of cheese, 2–3
Holes in cheese:
in blue cheese (piercing of), 104, 111 and *illus.*
in Swiss cheese, 73

Homogenized milk, 11–12, 14
Humidity, for aging cheese, 60
for bacterial and mold-ripened types, 103
for Swiss cheese, 73

Italian cheeses, 78
Mascarpone, 120
Montasio, 84–85; sharp variety, 85
Mozzarella, 78–80, *illus.* 79–80
Parmesan, 81–82; piquant, sharp variety, 82
Romano, 82–84; piquant, sharp variety, 84

Jalapeno peppers in cheese:
in Cheddar, 66
in Gouda, 71

Kefir cheese, 35–36
Knife, *illus.* 6, 7
for cutting curd, 52–53 and *illus.*, *illus.* 54

Labeling cheese, 60
Lactic acid, 15
Lactic cheese, 30–31, *illus.* 31–32
Lactose (milk sugar), 15
Ladling curds into cheesecloth, 15 and *illus.*
Leicester cheese, 61, 68–69
Lemon cheese, 14–15 and *illus.*, 36

Mascarpone, 120
Mat (for draining), 9 and *illus.*, 102, 103
Mesophilic starter culture, 16
for goat's milk cheeses, 93
how to prepare, 16–18 and *illus.*
Milk, 11–13
acidity of, *see* Acidity
goat's, 12, 93
homogenized, 11–12, 14
powdered, 14
raw, 12–13
pasteurizing (method), 13–14 and *illus.*
ripened, 16
ripening, *see* Ripening milk
skim, 12
whole, 11
Milk solids, 11
Milling curds, 56, *illus.* 57
Molding cheese:
bacterial and mold-ripened types, 102
hard cheeses, 57–58 and *illus.*
Mold powder:
blue, 103, 109
white, 96 and *illus.*, 103
Mold-ripened cheeses, 101
steps in making of, 101–104; piercing with holes, 104, 111 and *illus.*
varieties, *table* 102
for specific recipes see Blue mold-ripened cheeses; White mold-ripened cheeses
Molds (solid forms), 7 and *illus.*, 8, 57–58 and *illus.*
for Camembert cheese, *illus.* 104
for Coulommiers cheese, 106 and *illus.*
Montasio cheese, 84–85
sharp variety, 85
Mozzarella cheese, 78–80, *illus.* 79–80
Mysost cheese, 87–89

Neufchatel cheese, 36–37

Parmesan cheese, 80–82
piquant, sharp variety, 82
Pasteurizing (method), 13–14 and *illus.*
Pathogens, 12–13
Penicillium candidum (white mold powder), 96 and *illus.*, 103
Penicillium roquelorti (blue mold powder), 103, 109
Peppers in cheese:
in Cheddar, 66
in Gouda, 71
Piercing holes in blue cheeses, 104, 111 and *illus.*
Pot cheese, *see* Cottage cheese
Pots for cheesemaking, 6–7 and *illus.*
Powdered milk, 14
Presses, 58, *illus.* 59
cider press, 8, *illus.* 9
homemade, 8 and *illus.*
Wheeler type, 8 and *illus.*
Pressing cheese, 58–59 and *illus.*

Queso blanco, 29–30
Quick cheese, 119–120

Raw milk, 12–13
Record-keeping, 22
labeling cheese, 60
sample form, *illus.* 23
Red bacteria, 103, 115
for brick cheese, 115–17
Redressing cheese, 59
Refrigerator as "aging room," 60, 103
Rennet:
adding to ripened milk, 51 and *illus.*, 52
animal type, 19–20

Rennet: (continued)
for bacterial and mold-ripened cheeses, 101
coagulation method, 20–21
dilution of, 20–21 and *illus.,* 51
measuring, 21–22
and temperature, 21
vegetable type, 20
Ricotta cheese, 89–90
of goat's milk, 98–99
Ripening milk, 16, 50 and *illus.,* 51 and *illus.*
for bacterial and mold-ripened cheeses, 101
Roman (ancient) cheesemaking, 2
Romano cheese, 82–84
piquant, sharp variety, 84

Sage Cheddar cheese, 65–66
Saint Meure cheese, 96
Salt, 24
adding to milled curds, 56–57 and *illus.,* 103
see also Brine solution
Skim milk, 12
Slicing curds 52–53 and *illus., illus.* 54, 55
Soft cheeses, 27–45
varieties, *table* 27–28
for specific recipes see Bag cheeses; Cottage cheese; Cream cheese; Goat cheese; Quick cheese
Starter culture, 15–18
direct-set, 119
Starter culture, (continued)
mesophilic, 16, 50; for goat's milk cheeses, 93; how to prepare, 16–18 and *illus.*
storage of, 16, 17; freezing, 17–18 and *illus.*
thermophilic, 16, 50; for Emmenthaler cheese, 73, 76; how to prepare, 18; for Swiss cheese, 73
trouble-shooting, 18–19
varieties, 16
Sterilizing equipment, 5–6
Stilton cheese, 112–13
Stirring:
curds while cooking, 55 and *illus.;* with automatic stirrer, *illus.* 55
milk, from top, 52 and *illus.*
Suppliers of equipment (list), 129
Swiss cheeses, 73 and *illus.*
basic recipe, 74–76
caraway-flavored, 76
Emmenthaler, 76–77
Swiss cream cheese, 41

Temperature:
for aging cheese, 60; for mold-ripened cheeses, 103; for Swiss cheeses, 73
for coagulation, 21
for cooking curds, 55; for mold-ripened cheeses, 101; for Mozzarella, 78
and rennet 21
for ripening 50
Thermometer 6 and *illus.* 50
Thermophilic starter culture, 16, 50
for Emmenthaler cheese, 73, 76
Thermophilic starter culture, (continued)
how to prepare, 18
for Swiss cheese, 73
Top-stirring milk, 52 and *illus.*
Trouble-shooting, *table* 125–28
for starter culture, 18–19

Utensils, *see* Equipment

Varieties of cheese:
bacterial and mold-ripened, *table* 102
goat's milk, *table* 94
hard, *table* 48–49
soft, *table* 27–28
whey, *table* 88
Vinegar, 27

Washed curd cheeses, 69
Colby, 71–73
Gouda, 69–7 1; caraway-flavored, 71; with hot peppers, 71
Wax, 24, 60
Waxing cheese, 60 and *illus.*
Wheeler press, 8 and *illus.*
Whey, 11
components of, *table* 12
removal from curds, 53, 55–56
uses for, 87
Whey cheeses, 87
Gjetost, 89
Mysost, 87–89
Ricotta, 89–90
varieties, *table* 88
Ziegerkase, 90–91
White cheese, *see* Queso blanco

White mold powder, 96 and *illus.*, 103
White mold-ripened cheese:
Camembert, 104–106
Coulommiers, 106–107; English style, 107–108; French style, 108; herbed, 108
Saint Meure, 96
Whole milk, 11
Williams, Jesse, 3
Yogurt cheese, 34–35
Ziegerkase, 90–91

Other Storey/Garden Way Publishing Books You Will Enjoy